What Is Biodynamics?

What Is Biodynamics?

A Way to Heal and Revitalize the Earth

SEVEN LECTURES

Rudolf Steiner

INTRODUCTION BY HUGH J. COURTNEY

SteinerBooks

The lectures in this volume are from the following sources: "Spiritual Beings I" and "Spiritual Beings II" in *Spiritual Beings in the Heavenly Bodies and in the Kingdoms of Nature*, lectures 1 and 2, © Anthroposophic Press, 1992; "Elementals I" and "Elementals II" in *Harmony of the Creative Word*, lectures 7 and 8, revised translation by Matthew Barton, © Rudolf Steiner Press, 2001; "Agriculture I," "Agriculture II," and "Agriculture III" in *(Agriculture) Spiritual Foundations for the Renewal of Agriculture*, lectures 4, 5, and 6, translated by Catherine Creeger and Malcolm Gardner, editorial notes by Malcolm Gardner, © Bio-dynamic Farming and Gardening Association, Inc., 1993.

Introduction by Hugh J. Courtney © 2005
Illustrations by Matthew Baker © 2005

This edition has been edited by Marcia Merryman Means, who also wrote the short introductions before each lecture.

Published by SteinerBooks
610 Main Street, Great Barrington, MA 01230
www.steinerbooks.org

Library of Congress Cataloging-in-Publication Data is available.

10 9 8 7 6

Printed in the United States of America

Contents

So That the Earth May Be Healed: An Introduction
to Biodynamic Agriculture *by Hugh J. Courtney*   1

1. Spiritual Beings I   45
Helsinki, April 3, 1912   46

2. Spiritual Beings II   59
Helsinki, April 4, 1912   60

3. Elementals I   75
Dornach, November 2, 1923   76

4. Elementals II   91
Dornach, November 3, 1923   92

5. Agriculture I   106
Koberwitz, June 12, 1924   107

6. Agriculture II   134
Koberwitz, June 13, 1924   135

7. Agriculture III   160
Koberwitz, June 14, 1924   161

Further Reading   185

PUBLISHER'S NOTE

The lectures in this collection are taken from three different lecture cycles by Rudolf Steiner.

The first selection is from *Spiritual Beings in the Heavenly Bodies and the Kingdoms of Nature* (Helsinki, 1912), one of Rudolf Steiner's fundamental cycles. In it, he rises to the challenge of articulating in concrete detail the complex spiritual being underlying the phenomenal world. He does so by describing experientially the actions of the spiritual beings that work through the elements and forces of the Earth and the cosmos. At the same time, he demonstrates the role of humanity as the tenth spiritual hierarchy placed between the supra-earthly angelic beings "above" and the elemental and nature beings "below."

The second selection in our volume is from a lecture cycle ten years later, *Harmony of the Creative Word* (Dornach, 1923), where Steiner takes these insights further and shows how the threefold human being is a microcosm in which are reproduced the archetypes and formative tendencies of the cosmos. He describes the genesis of the kingdoms of nature out of the zodiac, emphasizing the dual process of the materialization of spirit and the spiritualization of matter through which evolution unfolds. Focusing on plants, birds, and butterflies, he shows the multiple ways in which metamorphosis occurs. Finally, he stresses the human function of becoming a cosmic artist, one who prolongs the work of the Heavens on Earth.

The last selection, from the lecture cycle generally called *The Agriculture Course* (Koberwitz, 1924), is the foundation of biodynamic agriculture. The course was given at Whitsun at the invitation of Count Carl Keyserlinck, who had asked Rudolf Steiner to come to his estate to address farmers about what could be done about the Earth, which was clearly under attack. About 130 people attended. Out of it, three tasks arose: to spiritualize the sciences; to sanctify the Earth; and to counteract the global danger threatening the Earth and her produce, including human food.

Each of these cycles is a *complete course* of lectures. Rudolf Steiner organized and thought through each cycle with a view to a coherent development of the ideas, practices, and concepts he wished to convey. Therefore, although individual lectures can be understood to some extent on their own, for a full and comprehensive understanding the entire course should be studied and individual lectures related to the totality. This is recommended for anyone seriously interested in pursuing the path to healing the Earth suggested in this book. Beyond that, it should be noted that each of the sources used here comes from a different time in Rudolf Steiner's life and was initially intended for a different audience. This explains certain differences of tone, language, and approach.

So That the Earth May Be Healed

An Introduction to Biodynamic Agriculture

by Hugh J. Courtney

When someone first stumbles onto biodynamic agriculture, it is fairly difficult to comprehend what it is about, since most proponents are hesitant to go into anything but the sketchiest of descriptions. Even then, those descriptions focus on the "farm as an individuality" or other intellectual concepts that still leave one somewhat in the dark as to how the actual practice of this form of agriculture is different from other approaches to agriculture. It is only with persistent questioning that the use of certain herbal preparations might be revealed. It is hoped that this introduction, with several carefully chosen lectures by Rudolf Steiner, will give clarity to the subject of biodynamic agriculture for the newcomer.

Rudolf Steiner (1861–1925) was a scientist and a philosopher, who initially earned a reputation as editor of the scientific works of Johann Wolfgang von Goethe. He could also be described as a seer—someone who could see into spiritual worlds—or in the popular vernacular, a clairvoyant. At the end of the nineteenth century, he began a twenty-five-year-long career of public and private lectures on esoteric subjects based on his explorations of spiritual worlds. Those explorations were done with the same scrupulous, exacting effort that he had used in his philosophical and scientific work up to that time. However, because the subject matter was beyond the recognition of the scientific establishment of his day, his esoteric research found an audience only among those open to matters concerned with the occult, which at that time, in particular, included members of the Theosophical Society founded by Madame Blavatsky. His acceptance among mainstream scientists evaporated and his work ever since has suffered a virtual silent treatment by the pillars of science and philosophy.

Not all the world, of course, ignored him and many in his day recognized the value of his research, which he called spiritual science or Anthroposophy. He also inspired the founding of Waldorf education, the art form of eurythmy and made valuable contributions to medicine, architecture, drama, and poetry, as well as many other fields of human endeavor. But it is his contribution to agriculture, called biodynamic agriculture, that may one day be recognized as perhaps his greatest gift to humanity. As for now, it remains virtually unknown in most of the world.

Steiner presented his course on agriculture in a series of eight lectures given in early June of 1924, less than a year before he died in March 1925.[1] For some time prior to his presentation of these lectures, a number of farmers, veterinarians, and others connected to the land had been urging him to provide them with his spiritual scientific insights into problems that were being encountered in agriculture. These problems included poor seed vitality, a reduction in the viability of forage crops that formerly could be counted on to last for years, increased breeding and disease problems among livestock, and greater concerns about plant disease. There was a considerable lapse of time, ostensibly because of his exceedingly busy lecture schedule, between the initial requests to Steiner to share his wisdom on this subject and the actual lectures. One can speculate that the long delay also occurred because Steiner had to conduct an enormous amount of spiritual scientific research into the realm of agriculture. During the lectures he finally gave on the subject, Steiner pointed out that a major source of the problems being experienced were due to the "dead" mineral fertilizers that had been increasingly used in farming since the time of Justus von Leibig.[2] Von Leibig and his fellow scientist, Mitscherlich proposed the so-called "Law of the Minimum"[3] in agriculture, which was instrumental in increasing dramatically the use of NPK (nitrogen/phosphorus/potash) fertilizers.

It is probable that Steiner's research for the Agriculture Course either revealed or further confirmed for him a number of facts that affected the realm of agriculture. Among those facts is the awareness that while the Earth is a living being, it is in a process of dying and the forces of nature are withdrawing from their previous role. A spiritually awakened humanity must take up the tasks that were originally the responsibility of such nature forces. However, humanity is not able to grow spiritually because the food that nourishes the human being no longer carries the forces that will lead to the necessary spiritual development. Steiner, in a conversation with Ehrenfried Pfeiffer discussed later, identified humanity's lack of spiritual development as "a problem of nutrition."[4] The ultimate solution to the problem of nutrition and the lack of spiritual development lies in finding the means to restore forces to nature that will then restore truly healthy life to the Earth. The means to do this were given by Steiner in the Agriculture Course, "so that the earth may be healed."[5] Through the indications Steiner gave, which are now known as the biodynamic agricultural preparations, he provided humanity with the means to accomplish the healing of the Earth.

Why is it that neither chemical nor organic agriculture can provide the means to re-enliven the Earth in the same way that is claimed for biodynamic agriculture? The challenge in answering this question is best met by identifying those characteristics within biodynamic agriculture that clearly distinguish it from other forms of agriculture. Whether one speaks of chemical, organic or traditional agriculture, there are three main factors within biodynamic agriculture that set it apart from these other more well-known forms of agriculture. In the first place, biodynamic agriculture approaches the farm as an organism or individuality in and of itself. Biodynamics arrives at this concept because it starts from the premise that the Earth itself is a living being. Second, the biodynamic farmer or gardener also attempts to relate his efforts to the movements not only of the Sun and Moon but of the other members of the solar system as well, all against a background of the entire cosmos. Finally, biodynamic agriculture also involves the use of nine very specially made herbal and mineral substances known as the biodynamic preparations.

One may object to the preceding statement regarding the biodynamic attitude toward the Earth as a living being by pointing out that traditional agriculture and even individual organic or chemical farmers may also have that attitude. While that may be true, the attitude in other forms of agriculture is confined to the realm of feeling, while the biodynamic farmer puts his feelings specifically into actual physical world practice through use of the biodynamic preparations.

Objections can be raised as well that farmers in traditional, organic and even chemical agriculture may take into account at least full and new moon rhythms in their everyday practice. Such practices are, however, based on a system of perceiving relationships within the heavens that existed some two thousand years ago, as will be explained later. The biodynamic farmer attempts to observe the physical reality of the heavens in a more exact and scientific way, and, in particular, may rely on other rhythms and relationships that go far beyond just the positions of the full and new moons. Often, the use of the nine biodynamic preparations is found to be more effective if applied in consonance with certain celestial rhythms and relationships. In fact, there is a distinct probability that use of the preparations serves to awaken certain energies among celestial relationships that remain dormant under organic, chemical or traditional agricultural practices.

Even if validity must be conceded to the objections cited on behalf of traditional, organic or chemical agriculture, one crystal clear difference remains exclusively to biodynamic agriculture—

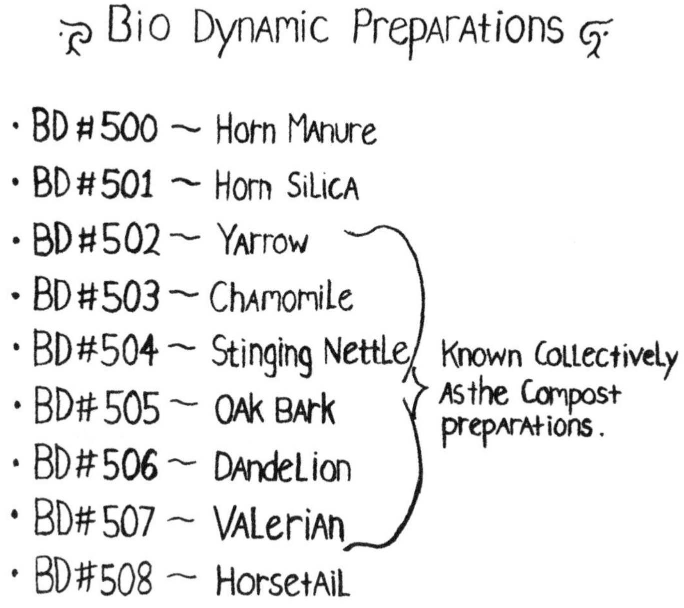

ॐ Bio Dynamic Preparations ॐ

- BD #500 ~ Horn Manure
- BD #501 ~ Horn Silica
- BD #502 ~ Yarrow
- BD #503 ~ Chamomile
- BD #504 ~ Stinging Nettle
- BD #505 ~ Oak Bark   } Known Collectively As the Compost preparations.
- BD #506 ~ Dandelion
- BD #507 ~ Valerian
- BD #508 ~ Horsetail

Figure 1.

the use of the nine biodynamic preparations. It is for this reason that this volume presents the three core lectures from Rudolf Steiner's Agriculture Course, which provide the indications for the nine biodynamic preparations basic to the practice of biodynamic agriculture. (Figure 1). It is appropriate now to make a brief introduction to these preparations before I go on to describe what I believe Steiner intended for them to accomplish in the world.

First of all, there are two spray preparations made in a cow horn. BD #500 is simply cow manure packed into a cow horn and buried in the Earth during the winter months. BD #501, at the opposite polarity in function, is ground silica put into a cow horn and buried in the summertime. The work of these two preparations is supplemented by a group of six preparations usually referred to as a "set" of biodynamic compost preparations. If a person first introduced to biodynamic agriculture is disturbed and upset by the use of cow horns, the manner of making some of the compost preparations will seem even more strange and astounding. BD #502, the yarrow preparation, is made by taking the blossoms of the yarrow plant and sewing them up inside a stag bladder, which then is hung in the summer sun and subsequently taken down and buried during the wintertime. BD #503, the chamomile preparation, involves using a bovine intestine, stuffing it full of chamomile blossoms and burying the resultant sausage in the Earth for the winter months also. BD #504, the stinging nettle preparation, is made by harvesting the entire above-ground plant at flowering time and burying it in the ground surrounded by peat moss not only for the winter months but through the following summer as well. BD #505, the oak bark preparation, uses ground oak bark and the fresh skull of a domestic animal from which all brain tissue has been removed, while retaining the meninges membrane lining the brain cavity. The bark is then packed into the empty skull. Once it is so prepared, the skull is buried for the winter in a swamp-like environment, such as a rain barrel full of vegetable matter. Explaining the method of production for this preparation is particularly difficult in any kind of public forum and is usually avoided by those representing biodynamics.

The next in the set of compost preparations is made by taking dandelion blossoms and sewing them up in a sack or pillow made from bovine mesentery or peritoneum membrane, which gives us BD #506, the dandelion preparation. It, too, is buried in the soil for the winter months. The last of the six compost preparations, BD #507, the valerian preparation, is simply the juice extracted from the blossoms of the valerian plant, which is then allowed to ferment for a few weeks before use. With the exception of the valerian, the finished product for all the compost preparations is transformed into a much more earthy and humus-like colloidal substance. This transformation is also most especially true of the BD #500 horn manure preparation.

When we come to the last of the preparations, BD #508, which is produced by making a tea from the horsetail herb,

Equisetum arvense, we are challenged by the fact that a good many biodynamic practitioners do not even regard it as a proper biodynamic preparation. Because it does not go through the kind of process involved in the making of the other preparations, not all of the present biodynamic community would include it as a bona fide preparation. Nevertheless, as will subsequently be addressed, I believe that it is of the utmost importance to the future of biodynamic agriculture that this preparation be considered a necessary part of biodynamic practice. The justification for that view to be provided later will, it is hoped, open up a broader perspective of the depths of meaning and purpose Rudolf Steiner intended for the preparations.

It is difficult for anyone who first comes to biodynamic agriculture to accept that such minuscule quantities of substance[6] as are used in practicing biodynamics have the tremendous effects that they do. However, when grapes are harvested in a cool wet September well ahead of the neighboring vineyards, because spray applications of BD #501 and BD #507 raised the Brix level (a measure of sugar content) one full point on the refractometer, acceptance becomes a little easier. Also, when over a period of a few years one creates topsoil down to a depth of fourteen inches, when a bare inch was present at the beginning of biodynamic treatment, the potency of the preparations becomes evident. Even more dramatic can be the arrival of blessed drought breaking rain when one applies in short order the nine biodynamic preparations in a "sequential spray" technique.[7] Although it is less dramatic, further evidence is presented by the continuing presence of a heavy dew on biodynamic farms or gardens when only drought-stressed land is seen otherwise in the neighborhood. Biodynamics cannot be grasped by intellects that are conditioned by an education that currently is so focused on the material world. Truly, biodynamics can be fully experienced only by applying all nine of the preparations. Real understanding takes place not just by exercising one's mental capacities, but only when one is "doing," or taking action.

When an anthology on a particular aspect of an author's work is extracted from the total body of that work, there are two clear caveats to be observed. This is especially so when dealing with such a diverse and comprehensive output as Rudolf Steiner's. In the first place, one is removing the chosen pieces, in this instance, several of Steiner's lectures, from the context of a particular lecture series, thereby inevitably affecting the depth of meaning to be gained. In the second place, the individual making the choice of lectures and attempting to introduce the reader to the subject must, of necessity, bring his or her own bias, educational background and prejudices to the presentation. Someone with a more thorough background in the sciences may well have chosen the first three lectures of Steiner's Agriculture Course, which provide an entirely new way of looking at chemistry and the periodic table of elements. They would also, perhaps, have salted their introduction with quotes from Steiner's many works dealing with the natural sciences. In this particular effort at an introduction, you will observe that there is a strong bias towards the actual biodynamic preparations, which are seen as a necessary part of practicing the art of biodynamic agriculture. There is an additional bias in favor of the unseen nature kingdom and the incredible spiritual powers that are deemed to be behind these biodynamic preparations.

One additional comment is in order here. In a very real sense, to study the Agriculture Course separate and apart from the entirety of Rudolf Steiner's work also results in taking that series of lectures out of context. To a very modest extent, the effort is being made here to place the Agriculture Course within the context of a certain aspect of Steiner's work. It is hoped that the effort will be received in the same light of reverence and awe with which its presentation is attempted.

In order to provide context for Steiner's agriculture lectures within the entire body of his work, and, in particular, to give the reader some sense of the incredible spiritual powers referred to previously, this introductory volume includes the first two lectures from another lecture cycle given in 1912, *Spiritual Beings in the Heavenly Bodies and in the Kingdoms of Nature*.[8] While these lectures were presented a number of years before the Agriculture Course, the background they offer on both the nature spirits and the spiritual beings behind them are of great value in understanding this aspect of the realm of agriculture. Within these lectures, Steiner suggests some seemingly simple exercises that, if diligently practiced, provide the possibility of acquiring a greater awareness of the elemental beings.

Also included in this volume are two additional lectures, the seventh and eighth, from the twelve lecture cycle, *Harmony of the Creative Word*,[9] on the subject of the nature spirits or elemental beings. Steiner presented this lecture cycle less than a year before giving the Agriculture Course. In fact, when it became known that Steiner was to give a lecture series on agriculture, it was assumed by many in the anthroposophical movement that such a course would address in much greater detail the role played by the nature spirits in specific agricultural tasks.[10] Only once in the course, however, as he speaks of the yarrow plant, does Steiner make reference to nature spirits just before he describes how one is to make the yarrow preparation.[11] Yet, it requires little imagination to see the lectures on the elemental beings as almost a necessary companion to the lectures on Agriculture. The preparations given in the Agriculture Course are intended to provide nourishment, in a very definite sense of the word, for the beings of the elemental kingdom. That nourishment is vitally necessary at this time in the world, and especially is it so if human evolution is to proceed in a positive manner. Without such nourishment, the nature spirits will remain estranged from humans, and will be indifferent to their part in the task of healing the earth.

One can make such statements in the hopes of encouraging a realization that biodynamic agriculture deserves greater attention; but this does not necessarily provide an answer to what sets biodynamic agriculture apart. While it is only since 1924 following Steiner's lectures on agriculture that this question has been in human consciousness, providing an answer has proved to be a substantial challenge. Among valiant written efforts available in English, we have the leaflet by Ehrenfried Pfeiffer titled *Bio-Dynamics—A Short Practical Introduction*;[12] the booklet *What Is Biodynamic Agriculture?*[13] by Herbert Koepf; as well as the essay, *What Is Biodynamics?*[14] by Sherry Wildfeuer. Wildfeuer's essay appeared in the year 2000 edition of *Stella Natura*, the annual biodynamic planting calendar.

One of the difficulties to be overcome in explaining this new form of agriculture to persons unfamiliar with the work of Rudolf Steiner is to put it into practical terms. However, of equal importance, is to answer this question from the spiritual side. To put this in another way, what is it that Steiner was really trying to convey when he finally presented these lectures after intense persuasion? What truly lies behind the message of biodynamic agriculture and, in particular, what lies behind the alchemical formulas now known as the biodynamic preparations? Before any attempt to present a possible answer to these questions, I will address the practical side of biodynamic agriculture.

The single most challenging aspect of introducing biodynamic agriculture is the common perception that it is merely organic agriculture with a peculiar twist. When trying to explain that this is not the case, many who attempt to present the subject often become apologetic or even defensive about biodynamic agriculture so as not to offend those practicing organic agriculture. On the other hand, our presentations are often quite "watered down" in the sense that we carefully avoid mentioning some of the more unusual aspects of biodynamic agriculture. Much emphasis is given to the more easily accepted ideas that are associated with biodynamic agriculture. Thus, such terms as "the farm as a closed or self-contained individuality" or "the farm as a living organism" serve as the introduction to the subject. References to the biodynamic agricultural preparations or to the use of a planting calendar are either almost totally avoided or given only the briefest mention with little elaboration.

Although biodynamic agriculture originated in 1924, the caution and even the protective secrecy that surrounded it in its beginnings, kept it well hidden from public view. When organic agriculture came along in the 1940s in the United States, its major proponent was J. I. Rodale, and its major organ of publicity was the magazine *Organic Gardening & Farming*.[15] An early contributor to that magazine was Ehrenfried Pfeiffer, who was the principal mentor of biodynamic agriculture in America. The strong personalities of both Rodale and Pfeiffer led to a parting of the ways in very short order. Thereafter, the elder Rodale showed no inclination to even recognize biodynamic agriculture in his publication. In the subsequent history of the magazine, those authors who chose to mention biodynamics may have even found it difficult for future articles to be accepted. Clearly this does not reflect an attitude supportive of such a form of agriculture. On a more positive note, various publications of Rodale Press in recent years are now using lengthy quotes from Pfeiffer himself to provide an explanation of the Biodynamic Method.[16]

Ultimately, one is forced to draw the conclusion that organic agriculture remains as firmly rooted in the material world as conventional or chemical agriculture. It is simply a matter of choosing more benign material substances to achieve one's goals of fertility and of avoiding the use of the kind of toxic materials that form much of the basis of chemical agriculture in today's world. Organic agriculture often must define itself by what it *avoids* or does *not* use. If one asserts that such a practical realm of human life as agriculture must be approached from a spiritual point of view, one may still be greeted with sneers and skepticism by organic as well as chemical farmers.

What is it that sets this spiritual form of biodynamic agriculture apart from both organic and conventional chemical agriculture? One significant factor is that biodynamic agriculture recognizes our planet Earth as a living being within a solar system and a greater universe that possess the quality of life as well. It is this that lies behind our references to the farm as "an individuality" or a self-contained organism. The fact that we as human beings now live in a world that makes no recognition that the Earth, our solar system, and the universe are all endowed with life is due to our worldview that only material nature is real.

In recognizing the living quality of the Earth, biodynamics also attempts to take into account the effect that the other living members of our solar system and the entire living cosmos have upon the Earth. We do this through the use of various biodynamic planting calendars. While their use is still very much at a beginning stage of research, many people have observed quite impressive results, which they attribute to the use of the calendars. It must be pointed out, however, that not all biodynamic practitioners adhere to the use of such calendars, having been unable to duplicate the results. Whether they used the same inputs or took into account all pertinent celestial factors are also questions not clearly answered.

As becomes readily apparent to someone using a biodynamic planting calendar, it bears little resemblance or correlation to the traditional farmers' almanacs, since it is based on the physical reality of the heavens, where the various members of the solar system are assigned to the constellation that actually lies behind them when viewed from the Earth. On the other hand, the tropical zodiac, on which the farmer's almanac and newspaper astrology columns are based, assumes that the Sun enters the first degree of the constellation/sign Aries/Ram at the time of the Spring equinox. The tropical zodiac further assumes that the constellations follow in order in nice even 30 degree segments to complete a 360 degree circle. Some two thousand years ago, during the time of Ptolemy, the topical zodiac and the sidereal zodiac coincided. Because of the Sun's apparent backward movement through the constellations, known as the precession of the equinoxes, the Sun's actual position is now somewhere in the early degrees of Pisces/Fishes, and soon, relatively speaking, will continue its backward movement to enter the constellation Aquarius/Waterman.

The most widely used biodynamic planting calendars are based on the astronomical system of constellation boundaries, although some calendars employ a sidereal zodiac which assigns each of the twelve constellations an exact 30 degree sector. The astronomical division results in some constellations having as much as a 43 degree sector while the narrowest constellation boundaries shrink to a mere 20 degrees. While celestial rhythms and their effect on agriculture may yet develop into a profound science, there is an enormous amount of research to be pursued over many years before it will have reached a plateau of certainty.

The most important factor that distinguishes biodynamic agriculture from organic or conventional chemical agriculture is the making and use of the biodynamic preparations. Biodynamic agriculture defines itself by the use of these preparations. It is only through them that the "life force," or biodynamic, energies are brought to bear in the organism of the farm or garden. In biodynamic agriculture the approach to fertility requires that we always remain within the realm of the living, which is accomplished by our use of the preparations.

Before tackling the deeper meanings lying hidden within biodynamic agriculture, let us take a closer look at the three works cited earlier and place them in historical context. The earliest of the works, by Pfeiffer, *Bio-Dynamics–A Short Practical Introduction* was first published in 1956. It totally avoids the use of the word spiritual. The initial paragraph merely identifies biodynamic farming and gardening as having originated out of the work of Rudolf Steiner "a philosopher known for his world-view called Anthroposophy (wisdom of man)."[17] Thereafter, a totally down-to-earth practical picture is drawn of a "method" of agriculture. There is very little in Pfeiffer's presentation of the principles and the practical steps needed to implement biodynamic agriculture with which the no-nonsense American farmer of the 1940s and 1950s would have had a problem. However the burying of cow horns filled with cow manure or crushed stones would have been a particularly hard sell. It is the manner of the making of the preparations above all else that led Pfeiffer to develop his proprietary formulas of the BD Compost Starter for making compost on a large scale and the BD Field Spray for general field use in agriculture. Those products gave hope of some degree of acceptance by mainstream agriculture.

Some twenty years later in 1976, the climate had changed somewhat when Dr. Herbert Koepf wrote the monograph, *What Is Biodynamic Agriculture?* Although this was published under the imprint of the Freie Hochschule fur Geisteswissenschaft Goetheanum (Free School for Spiritual Science at the Goetheanum), it has since been translated into English and, as a booklet, has served as an excellent introduction to put into the hands of someone seeking an answer to the question stated in the title. Dr. Koepf's introduction does not hesitate to describe the Agriculture Course as "spiritually-deepened knowledge."[18] Although the original audience for this work was quite familiar with Rudolf Steiner and Anthroposophy, by the year 1976 a more general audience could accept more readily the idea of a form of agriculture with a spiritual connection.

Koepf's work was not initially intended for the general public. The fact that it has now been translated and distributed beyond its originally intended audience, shows a certain amount of courage both on the part of the author and the publisher. While an explicit and detailed reference to the making of the preparations is not presented, both the substances and the sheaths involved are mentioned. Until this time, most publications on biodynamics had carefully avoided all but the sketchiest references to the preparations.

When we come to the briefer essay, "What Is Biodynamics," by Sherry Wildfeuer, we see further evidence of a much more accepting atmosphere for biodynamics as a spiritually connected form of agriculture. Wildfeuer does not hesitate to identify the fact that "we are strongly influenced by the unspoken philosophy of materialism which prevails in our education."[19] It is this philosophy of materialism that leads us "to accept actions towards the earth which are the consequences of a mechanistic view of living creatures and their mutual relations."[20] Counterposed to this view, is biodynamic agriculture, which "springs from a spiritual world view known as anthroposophy (from the Greek *anthropos*, meaning human being; *sophia*, wisdom)."[21] What anthroposophy through biodynamic agriculture can do, that goes beyond organic agriculture, is "to offer ways of healing the living earth through human insight into the formative forces of the cosmos."[22]

All of these works addressing the basic question concerning the nature of biodynamic agriculture give at least a hint that it has a spiritual side. As we have moved farther and farther away from the time the course was given in 1924, it appears that it has become easier and easier to refer to the spiritual foundation underpinning biodynamic agriculture. There is also a greater willingness to make reference to the source substances of the preparations and even to the sheaths used in making them. One notable omission in all three works is the lack of reference to the third of the biodynamic sprays, the horsetail herb, *Equisetum arvense*, or as it is referred to in at least some biodynamic circles, BD #508. For reasons to be elaborated upon later, the failure to promote this particular preparation to a place of equal importance among the others has been a contributing factor to the relatively insignificant impact of the biodynamic movement upon the world of agriculture.

Another major reason that the biodynamic effort has made so little impression upon mainstream agriculture is that we have lacked the courage to state clearly and unashamedly that this is a form of agriculture that depends upon the spirit. An additional underlying reason for the lack of progress is that we have not really learned to understand the tremendous power and force with which the several biodynamic preparations are endowed. In other words, we don't really believe in their efficacy ourselves, thereby making it difficult to convince others that such tiny particles of substance can actually change such huge areas of land to a condition of greater health.

While the efforts made by Pfeiffer, Koepf, and Wildfeuer are very helpful, in my own experience with the biodynamic preparations over the course of nearly thirty years, I have come to a much broader picture of Steiner's gift of biodynamic agriculture. That picture was developed as a result of my need to understand more fully Steiner's true intention in the Agriculture Course. The path I have taken to gain that understanding has focused on a fairly diligent reading of much of Steiner's work, as well as works by others who have studied Steiner's contributions to many fields of knowledge. After a certain time spent immersing myself in Steiner's body of wisdom, I came to realize that almost any lecture he ever gave contained at least a few sentences that served to shed light upon the meaning of one or another aspect of the agriculture lectures. I also realized that Steiner's words were always carefully chosen and that one needed to pay very close attention to what he says and how he says it.

Although the path I have followed in attempting to understand the agriculture lectures is a very personal one, by describing some of the signposts of that path, I may provide to others an insight into the deeper meaning of biodynamic agriculture. One of the first mysteries with which I was confronted, was trying to understand certain statements Steiner made to Ehrenfried Pfeiffer, which are described in what appears as the preface to the agriculture lectures in the translation by George Adams: "The most important thing is to make the benefits of our agricultural preparations available to the largest possible areas over the entire earth, so that the earth may be healed and the nutritive quality of its produce improved in every respect."[23]

This is followed very shortly by a second statement: "The benefits of the bio-dynamic compost preparations should be made available as quickly as possible to the largest possible areas of the entire earth for the earth's healing."[24] Then comes a third: "Spiritual scientific knowledge must have found its way into practical life by the middle of the century if untold damage to the health of man and nature is to be avoided."[25]

A mystery presented itself to me in these statements. Specifically, what was the reason for the urgency in using the biodynamic preparations? When that urgency is coupled with other concerns that Steiner expressed elsewhere regarding the need for spiritual science to have entered into the mainstream of civilization by the end of the century (that is, by the year 2000), one gets the impression that Steiner believed that the biodynamic preparations had a very important role to play in achieving the goal for spiritual science that Steiner had intended.

What in the world was Steiner trying to tell us when he gave us the biodynamic preparations? What was this danger that we were in, since the reality was that the preparations were not widely used by the end of the century? To answer this question calls for an enumeration of a few of Steiner's spiritual scientific concepts. One of those concepts is: that a plant is a being that lives between the Sun and the Earth and that one should regard a plant as though it were a human being whose head is planted in the Earth. (Figures 2 & 3) In viewing the plant we observe the organs of fruit and seed, flower, leaf and root, which correspond to the elements known in traditional wisdom as fire, air, water, and earth, respectively. This fourfold nature of the plant corresponds to the threefold nature of the human being that expresses itself in thinking, feeling and willing. The element/organ of earth/root correlates with the nerve/sense system in the human being. Similarly, the human rhythmic system encompasses what in the plant is water/leaf and air/flower. Finally, the human metabolic/limb system, which includes digestion and reproduction in the human being, relates to that which is fire/fruit (including seed) in the plant. In this picture we have the human being as a threefold entity.

Elsewhere in his considerable body of work, Steiner describes the human being as a four-fold being with a physical body, an etheric or life body, an astral or emotional body, and finally an ego body. When one looks at the traditional kingdoms of nature, that is, mineral, plant, animal and human, one can see that the

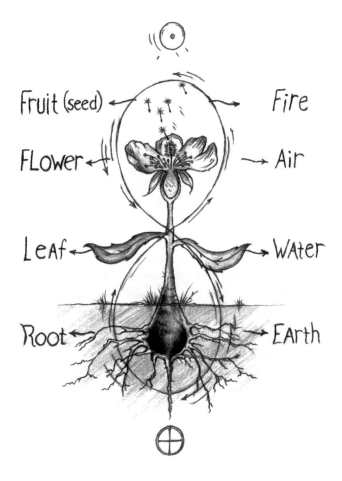

Figure 2.

ego body is exclusive to the human being. The human and animal kingdoms possess an astral body. The etheric body can be found in the human kingdom, the animal kingdom and the plant kingdom. The mineral kingdom possesses only a physical body.

Beyond this fourfold nature of the human being, Steiner ultimately describes the nine-fold nature of the human being: physical, etheric, and astral bodies; followed by a soul level of sentient

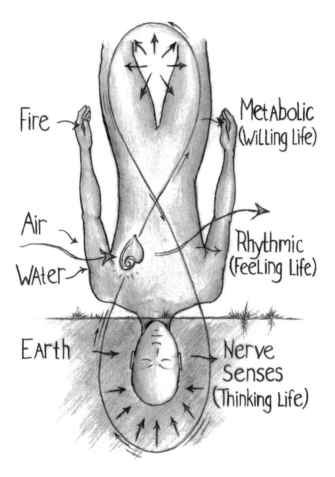

Fire

Air

Water

Earth

Metabolic
(Willing Life)

Rhythmic
(Feeling Life)

Nerve
Senses
(Thinking Life)

Figure 3.

soul, intellectual soul, consciousness soul; and finally the level of spirit, which includes spirit self, life spirit, and spirit human. The ultimate destiny of the human being is to fully develop in body, soul, and spirit. Later herein this ninefold nature of the human being will be related to the nine levels of subnature.

The elements of fire, air, water, and earth manifest through what Steiner referred to as the etheric formative forces. Those forces are

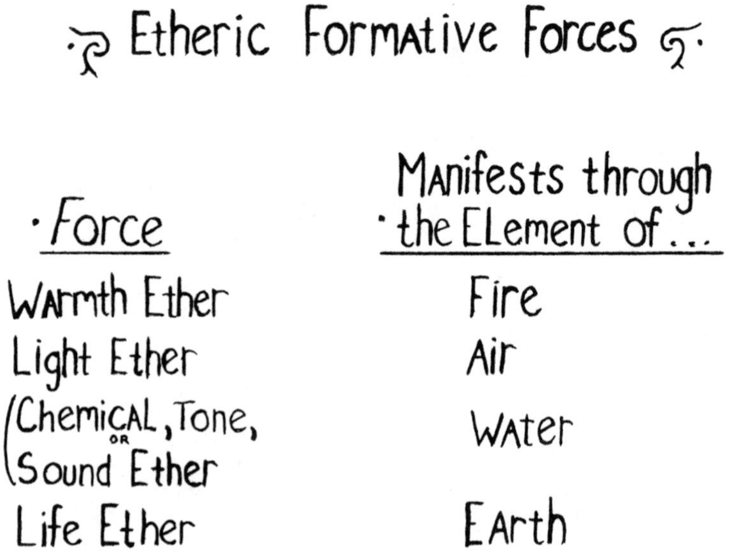

ॐ Etheric Formative Forces ॐ

· *Force*	Manifests through · the Element of...
Warmth Ether	Fire
Light Ether	Air
(Chemical, Tone, or Sound Ether	Water
Life Ether	Earth

Figure 4.

identified as warmth ether, light ether, chemical/tone/sound ether, and finally life ether (Figure 4). This particular concept, which Steiner developed through his spiritual scientific research, is helpful to consider when we explore the dynamic side of the biodynamic preparations. Steiner called for us to look at the biodynamic preparations as forces, not as substances. A great deal of help in understanding the preparations in this fashion can be found in the image that Steiner gives us in two lectures published under the title *Geographic Medicine*.[26] There Steiner describes the fact that the energies, or forces, in different parts of the world can be very different, even to the extent of being total polar opposites. Within agriculture limestone relates very strongly to the earthly nature or force, and silica relates powerfully to the cosmic nature or force, while clay serves as the mediating factor between those two polarities. Among the preparations, we also have a number of polarities involved, with certain preparations being more related to the

Bio-Dynamic Preparations Related to Geographic, Mineral, And other forces in the Earth
·ए ए·

Earthly	Mediating or Balancing	Cosmic
Western Hemisphere	Eastern Hemisphere	
North, Central, South America	Europe, Africa	Asia, Australia
Limestone	Clay/Humus	Silica
BD # 500	BD# 502 to507	BD# 508, BD#501

Figure 5.

limestone forces and therefore more earthly in nature, especially BD #500, horn manure. On the other hand, BD #501, horn silica, and BD #508, which can be called a plant silica (due to the high silica content in *Equisetum arvense*), are easily related to the silica force polarity as well as being more connected to the cosmic forces polarity. One can identify the various compost preparations as playing a part in the clay-mediating role with a greater or lesser relationship to either the limestone or silica polarity. When geographic forces are identified from Steiner's indications in *Geographic Medicine*, we can readily correlate the silica/cosmic tending preparations to the Eastern Hemisphere (Asia and Australia), the limestone/earthly preparation to the Western Hemisphere (North and South America), and the clay/mediating compost preparations to Europe (Africa, although not specifically mentioned by Steiner, may be paired with Europe). (Figure 5) This polarity picture presented by Steiner can be very useful in understanding some of the history of biodynamic agriculture and its development in various parts of the world. When one looks at this history, we find that what has developed in Europe is a very strong emphasis on just the clay-mediating factor involving the biodynamic compost preparations. The normal European approach to the implementation of biodynamic agriculture is to first make compost. We have the German biodynamic researcher, Maria Thun, developing her barrel compost recipe containing the six compost preparations. When Pfeiffer came to America from Europe he emphasized composting and ultimately developed the BD Compost Starter, because he was faithfully trying to carry out Steiner's mandate that the preparations be as widely used as possible.

On the other hand when one looks at Australia, one of the oldest areas of the eastern hemisphere, its strong cosmic energy nature can be readily seen. What developed in Australia under Alex Podolinsky, in particular, was a very strong emphasis on using regular BD #500, which was applied exclusively during the first two years of converting a farm to biodynamic practice. Thereafter, a "prepared" #500 (made of finished #500 with the compost preparations added) was used. It is claimed that in excess of two million acres in Australia are under this type of biodynamic care. We have in the geographic area of Australia, with its very strong cosmic energy orientation, a highly successful use of the earthly preparation BD #500. Thus we have a balancing of energies using an earthly preparation to counter the too-strong cosmic energies in Australia. If one takes that analogy and looks at the North and South American continents, you may come to the conclusion that the most important preparations for us to use are the silica polarity preparations so that cosmic forces can be brought to this very earthly western hemisphere. Unfortunately, that has not been the case, as we have allowed our biodynamic practice to be influenced first by the European emphasis on the compost preparations, and then by the Australian success with BD #500. The preparations we use least in America, BD #501 and BD #508, are exactly the ones we should be using most. As early as 1984, I could readily discern this imbalance with respect to the use of the preparations when I took up the activity of making and distributing preparations upon the death of the biodynamic pioneer in the United States, Josephine Porter. The major focus among practitioners was the use of the compost preparations in some form, either through the set of preparations themselves, or through the Pfeiffer BD Compost Starter (derived from the six compost preparations). A fairly large percentage also used BD #500, or as an alternative, the Pfeiffer BD Field Spray, which contained a very high proportion of BD #500. However, only a modest number of practitioners appeared to be using the BD #501. BD #508 was deemed appropriate for use only in the event of fungus conditions, since most practitioners had a very narrow interpretation of Steiner's indications regarding its use, that is, only as a counter to fungus problems caused by overly strong Moon forces.

In attempting to explore the reasons for the under use of BD #501 and BD #508, it turned out that well-respected biodynamic experts from Europe and Australia had cautioned against the use of BD #501, because they felt that the sun forces were already too strong in America and use of the silica could cause burning or scorching of a crop. My own observations suggest that this view requiring such cautious use of BD #501 does not necessarily apply in America to the same extent that has been experienced elsewhere in the world. This difference in point of view can be explained by the fact that these experts were taking their European or Australian experiences and applying them to conditions in America without any consideration that the geographic forces might require a different emphasis when using the biodynamic preparations.

As far as BD #508 (Horsetail herb) is concerned, one European practitioner took me to task quite vehemently, insisting that horsetail herb was not even a proper preparation and that under no circumstances should it be referred to as BD #508. My own picture of this particular preparation is woven from many different sources. In the first place, one should understand how Steiner structured many of his lectures. His practice was to state the problem to be solved; then near the end of a lecture, he would present the solution. In the main part of the lecture, he would build a picture of the many aspects of the problem. In the case of the sixth lecture in the Agriculture Course, he states the problem as a case of excessively strong Moon forces, which are behind weed, pest and especially fungal conditions. Near the end of the lecture, he provides the solution to curbing those forces by giving us the recipe for making a preparation from *Equisetum arvense*.

In another series of lectures, Steiner speaks of the Moon as the holder of the primal masculine force and of the comets as the carriers of the primal feminine force. It takes only a limited imagination to see the form of *Equisetum arvense* as an expression in the plant world of the same form exhibited by a comet in its repeating pattern of return (Figure 6). If one then takes a further imaginative step, one can understand that a primal feminine force is the one force that can substantially influence a primal masculine force. It requires little additional effort to picture the *Equisetum arvense* as being able not only to curb the strong moon forces, but to encourage those forces when they are too weak (as during drought conditions). In practice this is indeed the case.

Using the horsetail herb can actually attract or encourage rain a very significant percentage of the time, provided that it is used when the Moon is in a watery constellation (Cancer, Scorpio, or Pisces). Similarly, if horsetail is applied when the Moon is in a fire or warmth constellation (Aries, Leo, or Sagittarius), it appears that excessive moisture conditions can be curbed. While all this is very difficult to prove, one has only to try a few times the "sequential spray" technique. Using this spray during the time the Moon is in the appropriate constellation, one can experience directly that the horsetail herb is a very powerful tool to ameliorate or balance and harmonize the watery element. Such experience can provide the justification for labeling *Equisetum arvense* as BD # 508. One is then also justified in insisting that all *nine* of the biodynamic preparations be used and, most especially, that they be used here in the Western Hemisphere. Most important, the preparations

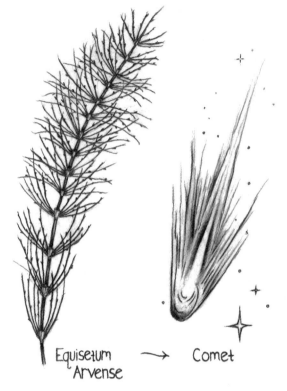

Figure 6.

need to be used as a totality rather than in isolation from one another, in order to achieve a harmony and balance of forces.

In *Geographic Medicine*, Steiner elaborates on the nature of energy in the various geographic areas of the Earth. He speaks of what are sometimes identified as the adversarial powers, Ahriman and Lucifer. Lucifer's domain in geographic terms is the Eastern Hemisphere, especially Asia, and Ahriman's domain is the Western Hemisphere. It is there that the energies are more of an earthly nature; or more connected to the being Ahriman and

associates, whose realm is that of subsensible nature, or that part
of nature lying below the level of sense perception. One encoun-
ters Ahrimanic forces in America much more strongly than any-
where else in the world. These Ahrimanic energies find their
expression in materialism and a total concern with the physical
world, and the effect of those forces is to pull the human being
down strongly into the material plane. The Luciferic forces or
beings, on the other hand, take one into illusory space. These
forces imbue the human being with an attitude of unconcern for
the physical world, which in its extreme form can result in behav-
ior that manifests as inability to cope with the practical aspects of
life. There are other adversarial powers, and Steiner speaks with
great concern about these entities, and about the fact that they
will manifest quite strongly at the end of the twentieth century to
further reinforce humanity's descent into materialism. These
adversarial powers, who are unalterably opposed to the Christ
impulse, are known as the Asuras, and specifically, the being
Sorath, the Sun Demon. This latter being is especially associated
with the occult number 666. Each time in history that that number or multipules of it arrive, the power of Sorath grows stronger.
That figure multiplied by three gives us 1998. That year, 1998,
was quite possibly the benchmark year that figured strongly in
Steiner's hope that spiritual scientific knowledge needed to manifest within the mainstream of civilization before the end of twentieth century. The presence of this Sun Demon is another
instance that underlines the necessity for the preparations to have
been widely used in the earth by the end of the last century.

Individuals such as Edgar Cayce, Gordon Michael Scallion,
and even Nostradamus along with many others, have foreseen
that the end of the twentieth century and the beginning of the
twenty-first would very likely see quite significant Earth changes
as the result of earthquakes, volcanic eruptions, and rising sea levels. It has been predicted that much of the West Coast of the
United States, for instance, would disappear beneath the sea.
Japan, China, much of Europe and many other parts of the world

will be affected by vast Earth changes. Someone with Rudolf Steiner's level of spiritual insight was undoubtedly able to tune into the probabilities of such future changes. Although he certainly did not dwell on such catastrophic possibilities, it is more than likely that, provided nothing intervened, Steiner foresaw an outcome that included future Earth changes. In anticipation and well beforehand, Steiner gave us the tools, in the nine biodynamic preparations, to counter the effects and ameliorate the consequences. It is my firm belief that these nine preparations can counter energies that might otherwise result in earthquakes and volcanic eruptions. The challenge lies in presenting a picture that might persuade you, the reader, of the enormous positive energies contained in the preparations.

In answering that challenge, an enormous debt is owed to a very interesting little journal with the intriguing name *Shoreline*, which had a very brief lifespan. In particular, the debt is owed to Issue no. 4 of that journal published in 1991, with the theme, "America, Sub-nature and the Second Coming." This theme was especially fascinating for me, because of my great interest in the destiny of America and my deep desire to fathom America's spiritual purpose. Please understand that reference is to an America without borders, not to just one country. To give you an intriguing picture of one aspect of that spiritual destiny, your attention is called to a particular technique in astrology and perhaps also used in astrosophy[27] which involves a projection of the celestial sphere upon the surface of the Earth. This technique has as its starting point a projection of the constellation Pleiades upon Egypt, particularly, the Plain of Giza, site of the Great Pyramid. When the other constellations are subsequently projected upon the Earth, the constellation Scutum, the Shield, and Aquila the Eagle land on the United States in North America. Aquila lies in the sector of the zodiac belonging to Scorpio, the Scorpion, which is associated in a positive future way with the mythical phoenix, the bird that rises from the ashes of destruction. It appears that Aquila also falls on at least part of Mexico. It is interesting to note that the national emblem of the United States is an eagle with a shield as breastplate and clutching an olive branch in one talon and a bundle of arrows in the other. The national emblem of Mexico is an eagle devouring a snake. These celestial energies are reflected in the geography of the continents and perhaps even the politics of the Earth as well. It is very interesting that Taurus, the Bull, falls upon the Taurus Mountains in Turkey and Leo, the Lion, on Great Britain, where the lion appears prominently in its national emblem or crest.

What, then, is the spiritual destiny of America in these times, when we are dealing with great adversarial powers as we move further into the new century? Let us look at a time two thousand years ago when humankind also had to deal with these adversarial powers. Steiner gives us a picture of those times in a lecture series on the Mexican mysteries.[28] In one of the articles in the *Shoreline* journal mentioned previously, there appears a fascinating essay based on Steiner's work, which summarizes the events that took place in Mexico. This article by Patrick Dixon is entitled "America, the Central Motif." Dixon describes the situation when the adversarial, or "dark," forces sought to counter what was taking place in the Middle East during the Christ experience. Those efforts to oppose what was transpiring in the Middle East were especially focused upon the crucifixion and the entire series of events Steiner refers to as the Mystery of Golgotha. These adversarial powers enlisted the services of a dark magician who was to achieve his ends by as much human sacrifice as possible. The methodology of that sacrifice was to gather all the members of a family or tribe, and one by one take and stretch the victim out over an altar, cut open the abdomen, reach up inside, pull out the heart, and perhaps, also, catch the blood to drink. All the while this sacrifice was taking place the other members of the family or tribe were compelled to watch, knowing that sooner or later they, too, would be sacrificed.

The purpose of this gruesome effort was to stimulate an enormous emotional outpouring of fear, and to slaughter the same number of human beings as had lived in the Jewish tribe from the time of Abraham through the lineage of Jesus of Nazareth. The outpouring of fear from the intended victims would so stimulate the elemental beings, or nature spirits, that in the depths of the Earth, at the level of subnature known as the fire earth level, they would cause enormous volcanic eruptions and earthquakes in order to balance out this outpouring of human fear. Steiner described the various realms of subnature in a series of early lectures with the title *At the Gates of Spiritual Science*.[29] There are nine levels (Figure 7) of sub-nature, which were labeled by Steiner as mineral earth, fluid earth, air earth, form earth, fruit earth, fire earth, earth mirror, earth severer, and earth core. These levels of subnature provide forces to oppose the previously indicated ninefold nature of the human being. If the dark magician could create sufficient fear by way of human sacrifice and project that fear down into the fire earth layer, the reaction of the elemental beings would result in numerous volcanic eruptions as well as earthquakes. The vast outpouring of smoke and debris from those volcanic eruptions, especially when they occurred in the middle latitudes of Mexico, would so obscure the Earth from the rest of the cosmos that those beings in the cosmos seeking to support the events connected with the Mystery of Golgotha would be cut off from rendering that support. The dark magician failed to prevent Christ's descent into hell and the resurrection, which were supported by the beings he sought to exclude from aiding the Christ event.

Steiner and Patrick Dixon tell us why he failed. He failed because of a white magician, Vitzliputzli. The means whereby Vitzliputzli was able to counter the evil forces and energies in the Earth was through music. He used the mountain chain of North America all the way through South America as a musical instrument. The singing and the music that Vitzliputzli and those around him created was of such a powerful and beautiful nature that it arrested the elemental beings; it weaned them away from the fear that was impacting their domain from human emotions experienced at the slaughter in Mexico. The elemental beings

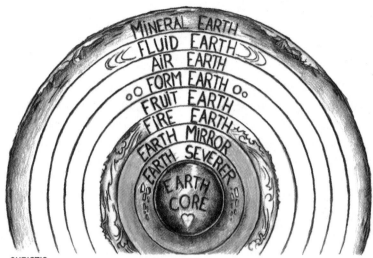

ꙷ INTERIOR OF THE EARTH ꙷ

CHRIST'S DECENT THROUGH THE EARTHLY SPHERES	SOURCES OF COUNTER IMPULSES TO:	DATES:
1. MINERAL EARTH	PHYSICAL BODY	MAR 1, 1933 to APR 1, 1945
2. FLUID EARTH	ETHERIC BODY	APR 1, 1945 to FEB 10, 1957
3. AIR EARTH	ASTRAL BODY	FEB 10, 1957 to DEC 21, 1968
4. FORM EARTH	SENTIENT SOUL	DEC 21,1968 to NOV 1, 1980
5. FRUIT EARTH	INTELLECTUAL SOUL	NOV 1, 1980 to SEPT 11, 1992
6. FIRE EARTH	CONSCIOUSNESS SOUL	SEPT 11, 1992 to JUL 23, 2004
7. EARTH MIRROR	SPIRIT SELF (MANAS)	JUL 23, 2004 to JUN 3, 2016
8. EARTH SEVERER	LIFE SPIRIT (BUDDHI)	JUN 3, 2016 to APR 14, 2028
9. EARTH CORE	SPIRIT MAN (ATMA)	APR 14, 2028 to FEB 24, 2040

FROM: *AT THE GATES OF SPIRITUAL SCIENCE*

ALTERNATIVE NAMES FOR LEVELS OF SUBNATURE FROM: *AN ESOTERIC COSMOLOGY*

1. MINERAL CRUST
2. NEGATIVE LIFE
3. INVERTED CONSCIOUSNESS
4. CIRCLE of FORMS
5. CIRCLE of GROWTH

6. CIRCLE of FIRE
7. CIRCLE of DECOMPOSITION
8. CIRCLE of FRAGMENTATION
9. EGO-CENTRIC-EGOISM

Figure 7.

were once again connected to what is referred to sometime "music of the spheres," the heavenly music that is so divi... that no one, especially an elemental being, can pay attention to anything else.

Please note one very important point here. Vitzliputzli used the etheric formative force that we now identify through Steiner as the chemical/tone/sound ether. This was some two thousand years ago, using the tone ether operating through the element water. The water in the air is what allows sound to be heard, and that allowed Vitzliputzli's music to reverberate throughout the entire mountain chain of the Americas.

Let us now take a look at another article in *Shoreline*, by Robert Powell, entitled "Sub-nature and the Second Coming." Robert Powell has spent a great deal of time and effort in trying to calculate the horoscope of the Christ event and the Christ being, using the indications Steiner gives in his many lectures on the subject, as well as the work of Anne Catherine Emmerich[30] to help determine the timing of events in the life of Christ. Powell reminds us of the description Steiner gives of the experience which every human being goes through at death: when, after completion of the life review which has been imprinted in our etheric body, the etheric body is dissolved. In one instance in human history, that etheric body did not dissolve, but remained whole. The risen Christ, the etheric body of the risen Christ, expanded to the outermost periphery of the cosmos and then it contracted. It expanded in both a twelve-year rhythm and a thirty-three and a third year rhythm. The twelve-year rhythm is associated with the number of years it takes for the planet Jupiter to complete one orbit around the sun through the constellations of the zodiac. Steiner spoke of the fact that there would be a reappearance of the Christ in the etheric beginning in 1933. Powell's (astrosophical) calculations establish that the etheric body or being of the Christ entered the realm of the mineral earth in approximately 1933, and after twelve years—by April 1, 1945—descended further into the realms of subnature (Figure 7). That twelve-year period is approximately the

time when we as spiritually conscious, awake, aware individuals
could have had an experience with this etheric Christ.

If we look at human history, April 1945 signals the advent of
the atomic bomb. The subsequent bombing of Hiroshima and
Nagasaki can be viewed as akin to the opening of the gates of hell
for the human kingdom, with humanity increasingly exposed to
the demonic influence of the subearthly spheres. In a certain sense,
a crucifixion of the etheric Christ took place when humankind
failed to recognize His presence, and the event we now call World
War II can readily be seen as analogous to that crucifixion. What
has transpired since, in a rhythm of twelve-years, is the further
descent of the being of Christ, the Self of Christ as Powell says,
through these various layers of the realm of subnature. Powell goes
on to say, "In each sub-earthly sphere a confrontation with the evil
at work in that sphere takes place—a confrontation that depends,
in the first place, on becoming conscious of the nature of the evil
principle at work."[31] Now Steiner speaks frequently about the
realms of supersensible nature, of the seraphim, the cherubim,
archangels, and angels, in fact of all the nine hierarchies of the
supersensible realm. In only a very few places does he speak of
these counterrealms, these mirror image or reverse image realms
belonging to the subnatural worlds. Now the domain of subnature
is specifically Ahriman's domain; while supersensible nature at
lower levels, is the domain where Lucifer holds sway.

During the period beginning September 11, 1992 through July
23, 2004, according to Powell's calculations, we received negative
energies from the fire earth realm of subnature, where the forces
counter to the consciousness soul can be found. It may be that
the adversarial powers were hopeful of stimulating sufficient fear
in human beings so that earthquakes and volcanic eruptions can
accomplish what was attempted two thousand years ago, that is
to cut off the Earth from the rest of the cosmos. The adversarial
powers may have jointly devised another way to screen human
beings from the cosmos, however. Picture what invisible radioac-
tivity has done to the Earth in weaving a radioactive shield against

the entry of cosmic energies. Picture what we have done in casting all sorts of human detritus into space to deflect cosmic energies. Picture all the invisible radiations of television, radio, electromagnetic and microwave transmissions, and look at the impact they have on humanity right now. Think of all the tremendous number of electrical power transmission lines around the earth. All this is akin to a massive spider web, both visible and invisible, which is forming to shield us from receiving cosmic energies needed to sustain life with spiritual depth on Earth.

What in the world do we have available as tools to counter these dreadful energies? We have none other than the nine biodynamic preparations given to us by Rudolf Steiner some eighty years ago. The biodynamic preparations heal not just earth substance; they heal the substance of water as well. They heal, albeit indirectly, the light and air and the warmth substances. By putting the biodynamic preparations on our farms and gardens, we can transform and cleanse the atmosphere by the daily inbreath and outbreath of the Earth, as that breath passes through our biodynamic soils. There might even be enough curative potency in that outbreath from our soils that it will have an impact on neighboring lands downwind when the next inbreath occurs.

Through these etheric formative forces and the preparations affecting them in a positive, curative way, we can accomplish a great deal of healing. If a person has ever observed, as I have, the phenomenon of the massive tic-tac-toe pattern of contrails in the sky, which takes place most often from midnight to four or five o'clock in the morning, one may wonder if this is not another means to shield the human being from cosmic energies. Ask those in authority about these contrails, and one finds that they have no answers. One can observe something very interesting about these contrails however, if one happens to live on a biodynamic farm. These cloud formations cannot hold their form; they break up and dissolve. That observation provides one small clue to the fact that we have a very great force in our hands in the biodynamic preparations, if we would but recognize it.

It would be useful to share with you some additional quotations from Steiner that were furnished by Powell in his article, "Sub-nature and the Second Coming." Powell says, "As a starting point, let us consider Rudolf Steiner's words from the night of the full moon, April 5-6, 1909, which he chose for the laying of the foundation stone of the Rosicrucian Temple at Malsch, near Karlsruhe, Germany."[32] Steiner says:

> We want to sink the foundation stone of this temple into the womb of our Mother Earth, beneath the rays of the Full Moon shining down upon us here, surrounded by the greenness of nature enveloping the building. And just as the Moon reflects the bright light of the Sun, so do we seek to mirror the light of the divine-spiritual beings. Full of trust we turn towards our great Mother Earth, who bears us and protects us so lovingly . . . In pain and suffering, our Mother Earth has become hardened. It is our mission to spiritualize her again, to redeem her, in that through the power of our hands we reshape her to become a spirit-filled work of art. May this stone be a first foundation stone for the redemption and transformation of our Planet Earth and may the power of this stone multiply itself a thousand fold.[33]

This was said in 1909, some time before Steiner gave us the biodynamic preparations for the spiritualization of the Earth. Human beings around Steiner at that time had to grow to a certain collective spiritual awareness before the agricultural preparations could be imparted to humanity. That Steiner was able to pass on this information before he died in 1925 was a most fortunate event. We must more fully put to use this vital information.

During the time when the Agriculture Course was being given, there were a number of other events taking place. Particularly meaningful was a conversation Steiner had at the end of the course with Countess Johanna von Keyserlingk, who was possibly one of the most advanced of Steiner's students. Powell translates that conversation as follows:

> Rudolf Steiner was good enough to come up to my room, where he spoke with me about the Kingdom in the interior of the Earth. We know that at the moment Christ's blood flowed onto the Earth a new sun-globe was born in the Earth's interior. My search had always been to penetrate the depths of the Earth, for I had seen raying up from there an Earth-core of gold, which Ptolemy designated as the "Archetypal Sun." I could not do otherwise, again and again, than to bring this golden ground into connection with the land of Shamballa, of which Rudolf Steiner had spoken. He had recounted how this land had disappeared from the sight of human beings and that Christ would open the door to those human beings seeking it, to lead them to the "sunken, fairy-tale land of Shamballa" of which the Hindus dream . . . I asked Rudolf Steiner, "Is the interior of the Earth of gold, originating from the empty space within the Sun, actually belonging back there again?" He replied, "Yes, the interior of the Earth is of gold." For the sake of certainty, I asked him further, "Herr doctor, If I stand here upon the ground, then beneath me, deep in the interior of the Earth, is the golden land. If I were to attain to freedom from sin and were to remain standing in the depths, the demons would not be able to harm me and I would be able to pass through them to the golden land. Is this so?" He replied, "If one passes through them together with Christ, then the demons are unable to harm one, but otherwise they would be able to destroy one!" He added the significant words, "However, they are able to become our helpers. Yes, this is so. The path is right, but it is very difficult!"[34]

Here we have Steiner speaking directly to Countess Keyserlingk about the realms of subnature, and of the elemental beings residing therein. Is it possible that Steiner could have foreseen the opening of the gates of hell that would unleash upon humankind the negative energies of these entities of the various levels of subnature? How can we recruit these demons to be our helpers? The means to do so is through none other than the nine biodynamic preparations, which Steiner had just imparted to humanity at Koberwitz. These tools are in our possession, and we need to use them.

Powell speaks of the relationship of the atomic bomb to the levels of subnature, to the opening of the mineral earth, to making contact with the fluid earth. According to Powell, the man-made sign of opening up to the air earth level, the third level of sub-nature, was Sputnik. Sputnik was an artificial satellite that could circle the Earth above the Earth's atmosphere, an event that marked the beginning of the space age. The deeper significance of this event was that man-made constructions began to be placed between the Earth and cosmic spiritual realms of existence, to block those cosmic energies. Powell identifies other historical events both positive and negative, such as the so-called sexual revolution, which included the spread of pornography; the civil rights movement and the fight against racial discrimination; peace demonstrations against the Vietnam War; the rise in terrorism, especially in the Middle East; and the tremendous increase in drug addiction. In particular Powell points out how the drug problem reached epidemic proportions with different drugs affecting the different levels of the human soul. How heroin injected into the blood overwhelms the sentient soul, and how cocaine overwhelms the intellectual soul. Powell goes on to point out that most of the signs that mark the deeper descent of Christ into a lower level of subnature originated in the United States— the destruction of Hiroshima with the atomic bomb; Star Wars, the popular label for the Strategic Defense Initiative, which intended to use space-based systems to protect against nuclear attack; and the widespread development of computers, for example. Only Sputnik originated elsewhere—in the Soviet Union—

and the United States quickly achieved preeminence over the Soviet Union in the space age. Powell goes on to say: "It is in the United States of America therefore that the most powerful links between man and sub-nature are evident"—as the geographic forces picture (Figure 5) bears out. He continues, "This applies not just to the technological side of man's alliance with the sub-earthly spheres but also to the cultural side."[35] The human experimentation and addiction vis-a-vis drugs and the pervasive influence of the "drug culture" on human society is a primary indication of that cultural alliance with the realms of subnature.

As of July 2004, we arrived at the close of the descent through the fire earth layer. This is the same fire earth level that Vitzliputzli dealt with at the time of Golgotha. It was about this level of sub-nature and its link with human passions that Steiner spoke of in *At the Gates of Spiritual Science:*

> Now, there are occasions when the very substance of the passions of the Fire-Earth begins to rebel. Aroused by men's passions, it penetrates through the Fruit-Earth, forces its way through the channels in the upper layers and even flows up into and violently shakes the solid Earth: the result is an earthquake. If this passion from the Fire-Earth thrusts up some of the Earth's substance, a volcano erupts.—There is still this connection between human passions and the passion layer in the interior of the Earth, and it is still an accumulation of evil passions and forces that gives rise to earthquakes and volcanic eruptions.[36]

Concerning the overcoming of bestial and depraved passions, Steiner says this:

> You will see that man is related to all the layers, for they are continually radiating out their forces. Humanity lives under the influence of these layers and has to overcome their powers. When human beings have learned to radiate

life on Earth and have trained their breathing so that it promotes life, they will have overcome the "Fire-Earth." When spiritually they overcome pain through serenity, they overcome the "Air-Earth." When concord reigns, the "Divisive" layer is conquered. When white magic triumphs, no evil remains on Earth. Human evolution thus implies a transformation of the Earth's interior.[37]

We have a great deal to look forward to and a tremendous amount of work ahead of us, work in which the biodynamic preparations can play a large role in overcoming the forces of negativity and counter-evolution emanating from the subearthly spheres. Powell suggests that another key that will allow us to counter the evil or noxious energies from these depths of subnature can be found in the nine beatitudes.[38] The sixth beatitude— "Blessed are the pure in heart: for they shall see God":—is the counter to the sixth level, through which the descent of the etheric Christ was to be completed by July 2004, according to Powell's calculations.

The Beatitudes to counter the seventh, eighth and ninth levels still awaiting the further descent of the etheric Christ are, respectively: "Blessed are the peacemakers for they shall be called the children of God," "Blessed are they which are persecuted for righteousness' sake: for their's is the kingdom of heaven." and "Blessed are ye, when men shall revile you, and persecute you, and shall say all manner of evil against you falsely, for my sake." How the further descent of the Christ will be resisted from these lowest levels can only be speculated about at this time, but remember, even that which has the appearance of evil can often be redeemed by the Christ energies as well as by our human action in service of the good.

Steiner spoke of the distinct possibility that even at the end of the twentieth century we could look forward to a manifestation of what he called the "War of All against All."[39] This occurs when the human ego is of such a nature that everyone is a kingdom unto oneself, taking no authority from anyone other than the self, and exhibiting no willingness to cooperate with others. It would seem that in many respects we are very close to this War of All against All on the Earth right now. Can we avoid it? There is hope that we can, but we must use these biodynamic preparations, which Steiner gave us to accomplish the task of converting the beings of subnature to helpers of the Christ.

Each of these preparations has an affinity for making smooth the way for Christ's descent and, ultimately, his ascent. When they are used, a vortexial energy channel is created, which allows the cosmic energies to penetrate to the core of the Earth from the outermost periphery of the cosmos. If only six of the preparations are used, as in making compost, the energy flow can descend only so far. Adding BD #500 to the mix gives a little further descent, and including BD #501 in practicing biodynamic agriculture takes us very far indeed. It is only with the addition of the ninth preparation, BD #508 (horsetail herb), that complete penetration of the cosmic energies is achieved, allowing those cosmic energies to reach the center of the Earth. That is the picture I have in my mind, having gone through a lengthy process studying Steiner and making and using the preparations to arrive at an understanding of them. If this picture is at all close to the mark, then even in a small backyard garden, there can be a significant positive effect, provided all nine preparations are used as intensively as possible. In this way we can bring down the cosmic energies to the elemental beings and awaken them into the service of the Christ.

To continue building a picture of the context surrounding the presentation of the Agriculture Course, there is another statement Steiner made during this period. In *Youth's Search in Nature*,[40] a lecture that Steiner gave to the young people attending the course, he tells them:

> To get to know the powers of nature beneath the earth leads to an understanding that the sword of Micha-el, the archangel Micha-el, as it is being forged, must be carried to

an altar that lies beneath the earth. There it must be found by receptive souls. It depends on your contribution for this sword of Micha-el to be found by more and more souls and it is not enough for it to be forged, something is really achieved only when it is found.

My imagination suggests that the tools we use to forge this sword of Micha-el are the biodynamic preparations.[41] The sword, when forged and laid in the earth is not only used by human beings; it is used by those elemental beings who take up this sword in the service of Christ.

One final quote from Rudolf Steiner will serve to underline the importance of what we can do with the preparations. In a statement from *The Christmas Mystery: Novalis the Seer*, Steiner speaks of a revelation that came to Novalis.[42] That revelation was: "The eye that beholds the Christ has itself been formed by the Christ-power. *The Christ-power within the eye beholds the Christ outside the eye.*"[43] Steiner reacts to this revelation of Novalis with the following statement: "These are truly powerful and mighty words! Novalis is also aware of the mighty truth that since the event of Golgotha the Being we call Christ has been the planetary spirit of the earth, the earthspirit by whom the earth's body will gradually be transformed."[44]

If humans had the consciousness of this inspiriting of the Earth by the Being of Christ, our desecration and pollution of the Earth would cease, and we would be more inclined to practice a spiritual husbandry toward the Earth made possible through the preparations. The attitude with which the Native Americans regarded the Earth was very much in keeping with the consciousness needed in our care for the Earth. This picture of the preparations is not a totally isolated view; Manfred Klett in *The Biodynamic Preparations as Sense Organs*, gives an image worth sharing:

I want to refer to one remark Rudolf Steiner made in a lecture in 1911. There he said "We will see in the course of this century and all the more in the future the decaying of the earth, of all our natural environment." He continues by saying that we will be very sorrowful about it and yet when we are really aware of this decaying world we will see that since the beginning of this century all of a sudden here and there something refreshing will spark from it. And finally he mentions those who are able to look into the background of this appearance can recognize new elemental beings which will become the servants of Christ.[45]

Klett uses the phrase "new." Indeed there may really be new elemental beings, but it is also possible that we will be witnessing the redemption of those elemental beings who have been forced through the indifference of human beings to serve Ahriman or other demonic beings. And they will be redeemed through this sword of Micha-el laid in the Earth through the use of the preparations. In speaking of these elemental beings, or new elemental beings, Klett continues:

They come into existence through a new relationship we seek to build to the decaying earth. With this in mind, imagine what we are doing in making and using the preparations. In making them we combine out of spiritual insight what is separate in nature. When using them we implant in nature a new material, the imposition of a new quality, which cannot be defined. The wisdom, which reigns in nature, is finite and somewhat defined by the ruling laws, but the quality of the matter we are now implanting is the most irrational. It is in its essence related to the quality of love, love is nothing but enlightened selfless will, the inauguratory substance of all future development. Love is the virtue to redeem, heal and transform all social distress. Why should we not be able to externalize this virtue and produce a force of irrational quality, which redeems, heals and transforms what is in a state of decay in nature. While working with the preparations I think we have to bear in mind this picture that we really do plant something into the earth that she longs for but cannot create for herself. It needs the free deed of man.[46]

To conclude, I would like to emphasize one last picture. Vitzliputzli accomplished his counterwork some two thousand years ago through the chemical/tone/sound ether force of his music. Our task at this time, during the descent and ascent of the etheric Christ, is to work via the force of the life ether, by always remaining within the realm of the living. We have no other tools to do that except the nine biodynamic preparations. We human beings, collectively and cooperatively, must become the Vitzliputzlis of this age through our use of the biodynamic preparations. It matters not whether we use them on the acreage of a huge farm or on a small backyard garden. What matters most is that all *nine* are used. Implant the sword of Micha-el in the earth through the preparations. Use them to summon the elementals to serve the Christ. Take up our human task of spiritualizing the Earth.

NOTES

1. There are now extant three versions in English as follows:

I. Steiner, Rudolf. *Agriculture–A Course of Eight Lectures*, (trsl. G. Adams, preface: E. Pfeiffer, Bio-Dynamic Agricultural Association, London, 1974. (Hereinafter: *Agriculture*).

II. Steiner, Rudolf. *(Agriculture) Spiritual Foundations for the Renewal of Agriculture*, (trsl. C. Creeger and M. Gardner, Editorial notes: M. Gardner, Bio-Dynamic Farming and Gardening Association, Inc., Kimberton, PA, 1993. (Hereinafter: *SFRA*).

III. Steiner, Rudolf, *Agriculture Course-The Birth of the Biodynamic Method*, (trsl. G. Adams, preface: E. Pfeiffer, Rudolf Steiner Press, Forest Row, England, 2004. (Hereinafter: *Agriculture–Birth of BD*)

Versions I and III are virtually identical except for the title difference. Version II is accompanied by copious footnotes as well as a facsimile reproduction and translation of Steiner's handwritten notes used in preparation for the lectures, and *Rudolf Steiner's Report to Members of the Anthroposophical Society after the Agriculture Course*. Either translation is valuable, and the serious student of biodynamic agriculture would be well-served to have both.

2. Baron Justus von Leibig (1803-1873) a German chemist who launched the era of chemical agricultural by his insistence that plants obtain new nutritive materials only through inorganic substances.

3. The "Law of the Minimum" basically stated that whatever essential inorganic plant nutrient was least available would dictate or limit the growth potential of the plant.

4. *Agriculture*, p.7. *SFRA* p. 261.

5. *Agriculture*, p.7. *SFRA* p. 260.

6. As little as a teaspoon of each of the compost preparations can be used on a pile of ten tons or more to accomplish the desired biodynamic transformation.

7. *Applied Biodynamics*, Issue no. 6, Winter, 1993. A quarterly publication of the Josephine Porter Institute for Applied Bio-dynamics, Woolwine, VA.

8. Steiner, Rudolf. *Spiritual Beings in the Heavenly Bodies and in the Kingdoms of Nature—A cycle of ten lectures.* Helsinki April 3-14, 1912. Anthroposophic Press, Hudson, NY, 1992.

9. Steiner, Rudolf. *Harmony of the Creative Word: The Human Being and the Elemental, Animal, Plant, and Mineral Kingdoms,* 12 lectures, Oct. 19-Nov. 11, 1923, Rudolf Steiner Press, Forest Row, 2001.

10. Klett, Manfred. *The Biodynamic Compost Preparations and the Biodynamic Preparations as Sense Organs—Talks on biodynamic agriculture.* International Biodynamic Initiative Group, Forest Row, England, 1996. p. 42.

11. *Agriculture*, p. 91-92. SFRA, p. 94.

12. Pfeiffer, Ehrenfried. *Bio-Dynamics—A Short, Practical Introduction*, Bio-Dynamic Farming and Gardening Association, Inc., Chester, NY, 1956. First published as an article in the journal *Bio-Dynamics*, Number 41, Autumn 1956, p. 2-9.

13. Koepf, Herbert. *What Is Biodynamic Agriculture?* Biodynamic Literature, Wyoming, RI, 1976. 43 pages.

14. Wildfeuer, Sherry. "What Is Biodynamics?" in *Stella Natura-2000*, Bio-Dynamic Farming & Gardening Association, Inc., San Francisco, CA, 1999. p. 4-6. This essay has also appeared for the past several years in an abridged form in the Biodynamic Resource Catalog, a publication of the Biodynamic Farming & Gardening Association, Inc.

15. This magazine has undergone numerous changes in name over the years. The current title is *Organic Gardening*, published by Rodale Press, Emmaus, PA.

16. Rodale, J. I. et al. *Encyclopedia of Organic Gardening.* Rodale Books, Inc. Emmaus, PA. Pp 87-89

17. Pfeiffer. *Biodynamics*, p.1.

18. Koepf. *What Is*, p.6.

19. Wildfeuer. In *Stella Natura-2000*, p. 4.

20. Wildfeuer. In *Stella Natura-2000*, p. 4.

21. Wildfeuer. In *Stella Natura-2000*, p. 4.

22. Wildfeuer. In *Stella Natura-2000*, p. 5.

23. *Agriculture*, p.7. *SFRA* p. 260.

24. *Agriculture*, p.8. *SFRA* p. 261.

25. *Agriculture*, p.11. Omitted in *SFRA*.

26. Steiner, Rudolf. *Geographic Medicine—Two lectures by Rudolf Steiner*, St. Gallen, Nov. 15 and 16, 1917. Mercury Press, Spring Valley, NY, 1986.

27. Astrosophy or the "wisdom" of the stars, as opposed to astrology, the "word" of the stars, arose out of the work of Willi Sucher and others who followed the indications given by Rudolf Steiner in arriving at an anthroposophical modernization of astrology.

28. Steiner, Rudolf, *Inner Impulses of Evolution—The Mexican Mysteries and the Knights Templar*, 7 lectures, Sept. 16-Oct. 1, 1916. Anthroposophic Press, Hudson, NY, 1984.

29. Steiner, Rudolf. *At the Gates of Spiritual Science*. 14 lectures, Aug 22 – Sept 4, 1906, Rudolf Steiner Press, London, 1986. These lectures also republished under the title *Foundations of Spiritual Science*.

30. Emmerich, Anne Catherine, was a 19th-century Catholic nun and mystic, who experienced numerous visions of biblical events all the way back to the creation of the world. Most of those visions centered on the life and ministry of Jesus Christ. Her visions were recorded by a German poet, Clemens Brentano. His journals were edited, mostly after his death, and subsequently published. The work is available in English in four volumes as *The Life of Jesus Christ and Biblical Revelations*, Tan Books and Publishers, Inc. Rockford, IL, 1986.

31. *Shoreline* no. 4, Living Art & Science Trust, Gwynedd, UK, 1991. p. 38. Issue No. 4 actually has two different sub-titles: *Subnature & the Parousia* and *America, Subnature & the Second Coming*.

32. *Shoreline* no. 4, p. 32.

33. *Shoreline* no. 4, p. 33.

34. *Shoreline* no. 4, p.34. But see also Keyserlingk, Adalbert, ed. *The Birth of a New Agriculture-Koberwitz 1924*. Temple Lodge, London, 1999. pp. 84-86.

35. *Shoreline* no. 4, p. 48.

36. Steiner. *At the Gates of Spiritual Science*. pp. 140-141.

37. Steiner. *At the Gates of Spiritual Science*. pp. 140.

38. *The Gospel According to Matthew*, Chapter 5, verses 3 – 11.

39. Steiner, Rudolf. *The Apocalypse of St. John*, Rudolf Steiner Press, London, 1977.

40. Steiner, Rudolf. *Youth's Search in Nature—A lecture given to the young people of Koberwitz*, June 17, 1924, Mercury Press, Spring Valley, NY, 1984. pp. 16-17.

41. The pronunciation emphasis on the last syllable is to indicate the Hebrew "el" meaning "of God."

42. Novalis, a German poet.

43. Steiner, Rudolf. *The Christmas Mystery—Novalis the Seer*, June 22, 1908, Mercury Press, Spring Valley, NY, 1985. p. 3.

44. *The Christmas Mystery*, p. 3.

45. *The Biodynamic Preparations as Sense Organs*, p. 72.

46. *The Biodynamic Preparations as Sense Organs*, pp. 72-73.

1

Spiritual Beings I

As HUMAN BEINGS, we exist in the physical world but also participate in the spiritual world. We experience the physical world through our senses and in our thinking and actions, but we can, if we make the effort, imbue these perceptions and actions with spiritual impulses. How can we find this spiritual connection? Spiritual science offers a schooling of the soul that can transform and deepen our feelings and perceptions and open the spiritual world to us. What do we find in that world of the spirit behind the natural world that exists around us, spread out in the myriad forms of the rocks and flowers, trees and animals, and human beings? We find multiple and various worlds of spirits. In the kingdoms of nature, we find an etheric life body of infinite diversity.

We can find our way into this world by means of meditative exercises that allow the sense impressions of nature to live vividly within us and then dissipate, leaving behind a certain soul devotion, a feeling of inner harmony with nature. The deeper aspect of nature behind ordinary sense impressions reveals itself to us then in the elemental beings. The first group of elementals, called gnomes, may be said to show themselves in solid forms to occult vision. They work deep within the warmth of the Earth in association with rocks and ores and bring the mineral element to the roots of plants, pushing them upward in their growth. The second group is the undines, whose forms are ever changing and who work aboveground in the moist, fluid atmosphere, drawing forth the plants from the ground and fostering their budding and leafing. Nature comes truly alive for us in spirit when we begin to recognize it as permeated with these and other elemental beings.

LECTURE BY RUDOLF STEINER[1]
APRIL 3, 1912

When our friends here gave me a warm invitation to come to them, they requested me to speak about the spiritual beings we find in the realms of nature and in the heavenly bodies. Our theme will compel us to touch upon a realm that is very far removed from all the knowledge humanity is given today by the external world, the intellectual world. From the very beginning we shall have to allude to a domain whose reality is denied by the outer world today. I shall take only one thing for granted, namely, that as a result of the studies you have made in spiritual science, you will meet me with a feeling and perception for the spiritual world; with respect to the manner in which we shall name things, we shall come to a mutual understanding in the course of the lectures. All the rest will, in certain respects, come of itself when, as time goes on, we acquire an understanding born of feeling and of perception for the fact that behind our sense world, behind the world that we as human beings experience, there lies a world of spirit, a spiritual world. And just as we penetrate into the physical world by regarding it not only as a great unity but also as separated into individual plants, animals,

---

1. Lecture 1 from *Spiritual Beings in the Heavenly Bodies and in the Kingdoms of Nature*, ten lectures, Helsinki, April 3–14, 1912 (GA 136).

minerals, peoples, and persons, so we can differentiate the spiritual world into distinct classes of individual spiritual beings. In spiritual science we do not speak merely of a spiritual world but instead of quite definite beings and forces standing behind our physical world.

What, then, do we include in the physical world? First, let us be clear about that. As belonging to the physical world, we reckon all that we can perceive with our senses: see with our eyes, hear with our ears, grasp with our hands. Further, we reckon as belonging to the physical world all that we can encompass with our thoughts insofar as these thoughts refer to external perception, to that which the physical world can say to us. In this physical world we must also include all that we, as human beings, do within it. It might easily make us pause and reflect when it is said that all that we as human beings do in the physical world forms part of that world, for we must admit that when we act in the physical world, we bring down the spiritual into that world. People do not act merely according to the suggestions of physical impulses and passions but also according to moral principles; our conduct, our actions, are influenced by morals. Certainly when we act morally, spiritual impulses play a part in our actions, but the field of action in which we act morally is nevertheless the physical world. Just as in our moral actions there is an interplay of spiritual impulses, even so do spiritual impulses permeate us through colors, sounds, warmth and cold—through all sense impressions. The spiritual is, in a sense, always hidden from external perception, from that which the external human being knows and can do. It is characteristic of the spiritual that we can recognize it only when we take the trouble, at least to a small extent, to become other than we have been hitherto.

We work together in our groups and gatherings. We hear there certain truths that tell us that there are various worlds, that the human being consists of various principles or bodies or whatever we like to call them, and that by allowing all this to influence us— although we may not always notice it—our souls will gradually

change to something different, even without our going through
esoteric development. What we learn through spiritual science
makes the soul different from what it was before. Compare your
feelings after you have taken part in the spiritual life of a working
group for a few years—the way you feel and think—with the
thoughts and feelings you had before or with the way in which
people think and feel who are not interested in spiritual science.
Spiritual science does not merely signify the acquisition of knowledge; it signifies most preeminently an education, a self-educa-
tion of our souls. We make ourselves different; we have other
interests. When one imbues oneself with spiritual science, the
habits of attention for this or that subject that one had developed
in previous years alter. What was once of interest is of interest no
longer; what was of no interest previously now begins to be interesting in the highest degree.

We ought not simply to say that only a person who has gone
through esoteric development can attain to a connection with the
spiritual world; esotericism does not begin with occult development. The moment that, with our whole heart, we make any link
with spiritual science, esotericism has already begun; our souls
begin at once to be transformed. There then develops in us something resembling what would arise, let us say, in a being who had
previously been able to see only light and darkness and who then,
through a special and different organization of the eyes, begins to
see colors. The whole world would appear different to such a
being. We need only observe it, we need only realize it, and we
shall soon see that the whole world starts to have a new aspect
when we have for a time gone through the self education we can
gain in a spiritual scientific circle. This self education to a perception of what lies behind physical facts is a fruit of the spiritual scientific movement in the world and is the most important part of
spiritual understanding. We should not believe that we can
acquire spiritual understanding by mere sentimentality, by simply
repeating continually that we wish to permeate all our feelings
with love. Other people, if they are good, wish to do that too.

This would only be giving way to a sort of pride. Rather, we should make it clear to ourselves how we can educate our feelings—by letting knowledge of the facts of a higher world influence us and by transforming our souls by means of this knowledge. This special manner of training the soul to a feeling for a higher world is what makes a spiritual scientist. Above all, we need this understanding if we intend to speak about the things that are to be spoken of in the course of these lectures.

One who, with trained occult vision, is able to see behind the physical facts, the physical world, immediately finds—behind all that is spread out as color, sound, warmth and cold and embodied in the laws of nature—beings that are not revealed to the external senses and the external intellect. Then, as one penetrates further and deeper, one discovers worlds with beings of an even higher order. If we wish to acquire an understanding of all that lies behind the sense world, then, in accordance with the special task that has been assigned to me here, we must take as our true starting point what we encounter immediately behind the sense world as soon as we raise the very first veil that sense perceptions spread over spiritual happenings. As a matter of fact, the world that reveals itself to trained occult vision as the one lying next to us presents the greatest surprise to contemporary understanding, to our present powers of comprehension. I am speaking to those who have, to some extent, accepted spiritual science. Therefore I may take it for granted that you know that immediately behind what meets us externally as human beings—behind what we see with our eyes, touch with our hands, and grasp with our understanding in ordinary human anatomy or physiology, behind what we call the physical body—we recognize the first supersensible human principle. This supersensible principle of the human being we call the etheric body, or life body.

Today we will not speak of still higher principles of human nature; we need be clear only that occult vision is able to see behind the physical body and to find there the etheric body, or life body. Occult vision can do something similar with regard to the nature around us. Just as we can investigate a human being occultly to see whether there is not something more than the physical body and then find the etheric body, so, too, we can look with occult vision at external nature in her colors, forms, sounds, and kingdoms—in the mineral, plant, animal, and human kingdoms, insofar as these meet us physically. We discover that just as behind the human physical body there is a life body, so there is a sort of etheric body, or life body, behind the whole of physical nature. But there is an immense difference between the etheric body of all of physical nature and that of the human being. When occult vision is directed to the human etheric, or life, body, it is seen as a unity, as a connected structure, as a single connected form or figure. When occult vision, on the other hand, penetrates all that external nature presents as color, form, or mineral, plant, or animal structures, it discovers that in physical nature the etheric body is a plurality, something infinitely multiform. That is the great difference; the human etheric, or life, body is a single unitary being, while there are many varied and differentiated beings behind physical nature.

Now I must show you how we arrive at such an assumption as that just made, namely, that behind physical nature there is an etheric, or life, body—strictly speaking, an etheric, or life, world, a multiplicity of differentiated beings. To express how we can arrive at this conclusion, I can clothe it in simple words: we are more and more able to recognize the etheric world, or life world, behind physical nature when we begin to have a moral perception of the world lying around us. What is meant by perceiving the whole world morally? What does this imply? First of all, looking away from the Earth, if we direct our gaze into the ranges of cosmic space, we are met by the blue sky. Suppose we do this on a day in which no cloud, not even the faintest silver white cloudlet, breaks the azure space of heaven. We look upward into this blue heaven spread out above us; it does not matter whether we recognize it in the physical sense as something real or not. The point is the impression that this wide stretch of the blue heavens makes upon us.

Suppose that we can yield ourselves up to this blue of the sky and that we do this with intensity for a long, long time. Imagine that we can do this in such a way that we forget everything else that we know in life and all that is around us. Suppose that we are able for one moment to forget all external impressions, all memories, all cares and troubles of life and that we can give ourselves completely to the single impression of the blue heavens. What I am now saying to you can be experienced by every human soul if only the soul will fulfill these necessary conditions; what I am telling you can be a common human experience. Suppose a human soul gazes in this way at nothing but the blue of the sky. A certain moment then comes, a moment in which the blue sky ceases to be blue, in which we no longer see anything that can in human language be called blue. If at that moment, when the blue ceases to be blue to us, we turn our attention to our own souls, we will notice quite a special mood within it. The blue disappears, and an infinity arises before us; in this infinity we experience a quite definite mood in our souls. A quite distinct feeling, a quite definite perception, pours itself into the emptiness that arises where the blue was before. If we would give a name to this soul perception, to what would soar out into infinite distances, there is only one word for it: it is a devout feeling, a feeling of pious devotion to infinity. All the religious feelings in the evolution of humanity have fundamentally a nuance that contains what I have here called pious devotion. The impression has called up a religious feeling, a moral perception. When within our souls the blue has disappeared, a moral perception of the external world springs to life.

Let us reflect upon another feeling by means of which we can attune ourselves in another way in moral harmony with external nature. When the trees are bursting into leaf and the meadows are filled with green, let us fix our gaze upon the green that in the most varied manner covers the Earth or meets us in the trees; again, we will do this in such a way as to forget all the external impressions that can affect our souls and simply devote ourselves to what in external nature meets us as green. If once more we can yield ourselves to what springs forth as the reality of green, we can carry this so far that the green disappears for us in the same way that previously the blue as blue disappeared. Here, again, we cannot say, "A color is spread out before our sight." Instead (and I remark expressly that I am telling you of things that each one of us can experience for ourselves if the requisite conditions are fulfilled), the soul has a peculiar feeling that can be articulated thus: "Now I understand what I experience when I think creatively, when a thought springs up in me, when an idea strikes me. I understand this now for the first time. I can learn this only from the bursting forth of the green all around me. I begin to understand the inmost parts of my soul through external nature when the outer natural impression has disappeared, and, in its place, a moral impression is left. The green of the plant tells me how I ought to feel within myself when my soul is blessed with the power to think thoughts, to cherish ideas." Once more an external impression of nature is transmuted into a moral feeling.

Or we may look at a wide stretch of white snow. In the same way as the blue of the sky and the green of Earth's robe of vegetation, so, too, can the white of the snow set free within us a moral feeling for all that we call the phenomena of matter in the world. If, in contemplation of the white snow mantle, we can forget everything else and experience the whiteness and then allow it to disappear, we obtain an understanding of that which fills the Earth as substance, as matter. We feel matter living and weaving in the world. Just as one can transform all external sight impressions into moral perceptions, so can one transform impressions of sound into moral perceptions. Suppose we listen to a tone and then to its octave and so attune our souls to this dual sound of a tonic note and its octave that we forget everything else, eliminate all the rest, and completely yield ourselves to these tones. It then comes about at last that instead of hearing these dual tones, our attention is directed away from them, and we no longer hear them. We find once again that in our souls a moral feeling is set free. We begin to have a spiritual understanding of what we experience when a wish lives within us that tries to lead us to something and then our reason influences our wish. The concord of wish and reason, of thought and desire, as they live in the human soul is perceived in the tone and its octave.

In like manner we might let the most varied sense perceptions work upon us; we could in this way permit all that we perceive in nature through our senses to disappear, so that this sense veil is removed; then moral perceptions of sympathy and antipathy would arise everywhere. If we accustom ourselves in this way to eliminate all that we see with our eyes or hear with our ears, that our hands grasp, or that our understanding (which is connected with the brain) comprehends—if we eliminate all that and accustom ourselves nevertheless to stand before the world—then there works within us something deeper than the power of vision of our eyes or the power of hearing of our ears or the intellectual power of our brain thinking. We confront a deeper being of the external world. The immensity of infinity so works upon us that we become imbued with a religious mood. The green mantle of plants works upon us such that we feel and perceive in our inner being something spiritually bursting forth into bloom. The white robe of snow works upon us such that we gain an understanding of the nature of matter, of substance, in the world; we grasp the world through something deeper within us than we had hitherto brought into play. In this way we come into contact with a deeper aspect of the world itself. The external veil of nature is drawn aside, as it were, and we enter a world that lies behind this external veil. Just as when we look behind the human physical body we come to the etheric body, or life body, so in this way we enter a region in which, gradually, manifold beings disclose themselves—those beings that live and work behind the mineral kingdom. The etheric world little by little appears before us, differentiated in its details.

In occult science, what gradually appears before us in the way described has always been called the elemental world; those spiritual beings we meet there, of whom we have spoken, are the elemental spirits that lie hidden behind all that constitutes the physical sense-perceptible world. I have already said that whereas the human etheric body is a unity, the etheric world of nature is a plurality, a multiplicity. Since we perceive something quite new, how can we find it possible to describe that which gradually impresses itself upon us from behind external nature? We can do so if, by way of comparison, we make a connecting link with what is known. In the whole array that lies behind the physical world, we first find beings that present self-enclosed pictures to occult vision. To characterize what we first find there, I must refer to something already known. We perceive self enclosed pictures, beings with definite outlines; we can say that they can be described according to their forms, or shapes. These beings are one class that we find behind the sense world. A second class of beings we can describe only if we turn away from what shows itself in a specific form, with a set figure, and employ the word metamorphosis or transformation. That is the second phenomenon that presents itself to occult vision. Beings that have definite forms belong to one class. Beings that change their shapes at every moment and who, as soon as we meet them and think we have grasped them, immediately transform into something else, so that we can follow them only if we make our souls mobile and receptive, such beings belong to the second class. Occult vision actually finds only the first class of beings, those that have quite a definite form, when (starting from such conditions as have already been described) it penetrates into the depths of the Earth.

I have said that we must allow all that works on us in the external world to arouse a moral effect. By way of example, we have brought forward the ways in which one can raise the blue of the heavens, the green of the plants, the whiteness of the snow into moral impressions. Let us now suppose that we penetrate into the interior of the Earth. When, let us say, we associate with miners, we reach the inner portion of the Earth; at any rate, we enter regions in which we cannot at first so school our eyes that our vision is transformed into a moral impression. But in our feeling we notice warmth, differentiated degrees of warmth. We must first feel the warmth; warmth must be the physical impression of nature when we plunge into the realm of the Earth. If we keep in view these differences of warmth, these alternations of temperature, and all that otherwise works on our senses because we are underground and if we allow all this to work upon us, then through thus penetrating into the inner part of the Earth and feeling ourselves united with what is active there, we go through a definite experience. If we then disregard everything that produces an impression; if we exert ourselves while we are down there to feel nothing, not even the differences of warmth that were only for us a preparatory stage; if we try to see nothing, to hear nothing, but to let the impression so affect us that something moral issues from our souls—there arises before our occult vision those creative nature beings who, for the occultist, are really active in everything belonging to the Earth, especially in everything of the nature of metal, and who now present themselves to the imagination, to imaginative knowledge, in sharply defined forms of the most varied kind.

If, having had an occult training and having at the same time a certain love of such things—it is especially important to have this here—we make acquaintance with miners and go down into the mines and if we can forget all external impressions when we are down there, we will then feel rising up before our imagination the first class of beings that create and weave behind all that is earthy and especially in all that pertains to metals. I have not yet spoken today of the way in which popular fairy tales and folk legends have made use of all that, in a sense, actually exists; I should like first to give you the dry facts that offer themselves to occult vision. For according to the task set me, I must first go to work empirically—that is, I must give an account, first of all, of what we find in the various kingdoms of nature. This is how I understand the topic put before me.

Because we must think of the individual human being as permeated by what appears to occult sight as the etheric body, so must we think of all that is living and weaving in the world as permeated by a multiplicity of spiritual living forces and beings. Just as, in occult vision, we perceive in our imagination clearly outlined nature beings—and in this way can have before us beings with settled forms, of whom we see outlines that we could sketch—so it is also possible for occult vision to have an impression of other beings standing immediately behind the veil of nature. Let us say we have a day when the weather conditions are constantly changing—when, for instance, clouds form and rain falls and when, perhaps, a mist rises from the surface of the Earth. If on such a day we yield to such phenomena in the way already described, so that we allow a moral feeling to take the place of a physical one, then we may again have quite a distinct experience.

This is especially the case if we devote ourselves to the peculiar play of a body of water tossing in a waterfall and giving out clouds of spray, if we yield ourselves to the forming and dissolving mist and to the watery vapor filling the air and rising like smoke or if we see the fine rain coming down or feel a slight drizzle in the air. If we feel all this morally, there appears a second class of beings, to whom we can apply the word metamorphosis or transformation. As little as we can paint lightning can we draw this second class of beings. We can note only a shape, present for a moment, and the next moment everything changes. Thus there appear to us as the second class those that constantly change form, for which we can find a symbol for the imagination in the changing formations of clouds.

As occultists we become acquainted in yet another way with these beings. When we observe the plants as they come forth from the Earth in springtime, just when they put forth the first green shoots—not later, when they are getting ready to bear fruit—the occultist perceives that those same beings which he discovered in the pulverizing, drifting, gathering vapors are surrounding and bathing the beings of the budding plants. We can say, then, that when we observe plants springing forth from the Earth, we see them everywhere bathed by such ever-changing beings as these. Then occult vision feels that that which weaves and hovers unseen over the buds of the plants is in some way concerned with what makes the plants push up out of the ground, drawing them forth from the ground. You see, ordinary physical science recognizes only the growth of the plants and knows only that the plants have an impelling power that forces them upward from below. The occultist, however, recognizes more than this in the case of the blossom. The occultist recognizes around the young sprouting plant changing, transforming beings that have been released from the surrounding space and penetrate downward. They do not, like the physical principle of growth, merely pass from below upward but instead come from above downward, and draw forth the plants from the ground. In spring, when the Earth is robing herself in green, to the occultist it is as though nature forces, descending from the universe, draw forth that which is within the Earth, so that the inner part of the Earth may become visible to the outer surrounding world, to the heavens.

Something that is in unceasing motion hovers over the plant, and occult vision acquires a feeling that it is the same as what is present in rarefied water, tossing itself into vapor and rain: the second class of nature forces and nature beings. In the next lecture we shall pass on to the description of the third and fourth classes, which are much more interesting, and then all this will become clearer. When we set about making observations such as these, which lie so far from present human consciousness, we must always bear in mind that all that meets us is physical but permeated by the spiritual.

The course to be followed in our considerations shall be such that we shall first describe simply the facts that an occultly trained vision can experience in the external world, facts that are evident to us when we look into the depths of the Earth or the atmosphere, into what happens in the different realms of nature and in the heavenly spaces filled by the fixed stars. Only at the end shall we gather the whole together in a kind of theoretical knowledge that is able to enlighten us as to what lies as spirit at the foundations of our physical universe and its different realms and kingdoms.

2

Spiritual Beings II

WE FIND THE THIRD GROUP OF ELEMENTALS, the sylphs, at work in the withering and fading of mature plants. These nature beings live in the air, played upon by the light and warmth of the Sun. They glitter and spark, arising here and vanishing there. When plants come to fruition and then go to seed, the fourth group of elemental beings, once given the name salamanders, comes into its own. Salamanders, or fire spirits, live in warmth and protect the plant germ, carrying it from one season of growth into the next and fostering its ripening by imbuing it with the heat it needs to develop. Here, then, we have the nature beings of earth, water, air, and fire, which together form and indeed create, in the kingdoms of nature, the etheric life body of our planet.

Further meditative exercises allow us to delve still deeper into the spirit world and find those beings that make up the Earth's astral body. The Spirits of the Cycles of Time, as they are called, govern the rhythmic, regular succession of day into night and night into day again, one season into the next, as well as the spinning of the Earth itself on its axis. In their governance of the laws of nature, these spirits oversee the activities of the nature beings.

Further esoteric development reveals to us a still-higher level of the spirit world, the planetary spirit, which brings Earth into relationship and harmony with the other heavenly bodies. Seeking for the deeper purposes behind the natural phenomena that greet us everywhere in ordinary life, we approach this planetary spirit and find there what unites us with the cosmos and gives meaning to our world.

Lecture by Rudolf Steiner[1]
April 4, 1912

Yesterday, I tried first of all to point the way that leads the human soul to the observation of the spiritual world hidden directly behind our material physical world and then to draw attention to two classes or categories of spiritual beings, perceptible to occult vision when the veil of the sense world has been drawn aside. Today, we shall speak of two other forms or categories of nature spirits. The one is disclosed to trained occult sight when we observe the gradual fading and dying of the plant world in late summer or autumn, the dying of nature beings in general. As soon as a plant begins to develop fruit in the blossom, we can allow this fruit to work upon the soul in the manner described in our last lecture. In this way, we receive in our imagination the impression of the spiritual beings concerned with the fading and dying of the beings of nature.

We were able to describe yesterday that in spring the plants are, so to speak, drawn out of the Earth by certain beings that are subject to perpetual metamorphosis. Likewise we can say that when, for instance, plants have finished this development and the time has come for them to fade, other beings then work upon them— beings of whom we cannot even say that they, too, continually

---

1. Lecture 2 from *Spiritual Beings in the Heavenly Bodies and in the Kingdoms of Nature.*

change their forms, for, strictly speaking, they have no form of their own at all. They appear flashing up like lightning, like little meteors—now flashing up and now disappearing. They really have no definite form but flit over our Earth, flashing and vanishing like little meteors or will-o'-the-wisps.

These beings are primarily connected with the ripening of everything in the kingdoms of nature; the ripening process comes about because these forces or beings exist. They are visible to occult vision only when it concentrates on the air itself, indeed, on the purest air possible. We have described the second sort of nature beings—those that draw the plants out of the ground in spring—by saying that to perceive them we must allow falling water or water condensed into cloud formations or something of a like character to work upon us. Now air, as free from moisture as possible and played upon by the light and warmth of the Sun, must work upon the soul if we are to visualize in our imagination the third group, the meteor-like flashing and disappearing beings that live in air free from moisture and eagerly drink in the light that permeates the air and causes them to flash and shine. These beings then sink down into the plant world or the animal world and bring about their ripening and maturity. In the very way we approach these beings, we see that they stand in a certain relation to what occultism has always called the elements.

We find what we described in the last lecture as the first class of such beings when we descend into the depths of the Earth and penetrate the solid substance of our planet. Our imagination is then confronted with beings of a definite form, and we may call these the nature spirits of solid substance, or the nature spirits of the Earth. The second category are to be found in water that collects and disperses, so that we may connect these spiritual beings with what in occultism has always been called the fluid, or watery, element. In this element they undergo metamorphoses, and at the same time they do the work of drawing forth from the Earth everything that grows and sprouts. The third group—the flashing, disappearing beings—stands not in connection with the watery element but in connection with the element of air, air when it is as free from moisture as possible. Thus we may now speak of nature spirits of earth, water, and air.

There is a fourth category of such spiritual beings with which occult vision can become familiar. For this, occult vision must wait until a blossom has brought forth fruit and seed and then take note of how the germ gradually grows into a new plant. Only then can such vision be achieved with ease—in other conditions it is difficult to detect this fourth kind of being, for they are the protectors of the germs, of all the seeds in our kingdoms of nature. As guardians, these beings carry the seed from one generation of plants or other nature beings to the next. We can observe that these beings, the protectors of the seeds or germs, make it possible for the same beings to reappear continually on our Earth and for these beings to be brought into contact with the warmth of our planet—with what from early times has been called the element of fire or heat. That is why the forces of the seed are also connected with a certain degree of heat, a specific temperature. If occult vision sees accurately, it finds that the necessary transmutation of the warmth of the environment into such heat as is needed by the seed or germ in order to ripen—that the changing of lifeless warmth into a living heat—is provided for by these beings. Hence, they can also be called the nature spirits of fire, or of heat. To begin with, then, we have become acquainted with four categories of nature spirits, having a certain relation to what are called the elements of earth, water, air, and fire.

It is as though these spiritual beings had these elements as their jurisdiction, their territory, just as we human beings have as ours the whole planet. Just as the planet Earth is our home in the universe, so these beings have their home in one or another of the elements mentioned. We have already drawn attention to the fact that for our earthly physical world, that is, for the Earth as a whole, with its various kingdoms of nature, these different beings signify what the etheric body, or life body, signifies for individual human beings. We have said, however, that in the human being this life body is a unity, whereas the etheric body of the Earth consists of many, many such nature spirits, which are, moreover, divided into four classes. The living cooperation of these nature spirits constitutes the etheric body, or life body, of the Earth. Thus, it is not a unity but a multiplicity, a plurality. If we wish to discern this etheric body of the Earth with occult vision, then we must allow the physical world to influence us morally, thereby drawing aside the veil of the physical world. Then the etheric body of the Earth, which lies directly behind this veil, becomes visible.

What do we find when we draw aside this further veil, described as the etheric body of the Earth? We know that behind the etheric body is the astral body, the third principle of the human being—the body that is the bearer of desires, wishes, and passions. If we disregard the higher principles of human nature, we may say that in a human being we have first of all a physical body, behind which is an etheric body and behind that an astral body. It is just the same in external nature. If we draw aside the physical, we certainly come to a plurality. This represents the etheric body of the whole Earth, with all its kingdoms of nature. Can we also speak of a sort of astral body of the Earth, something that, in relation to the whole Earth and to all its kingdoms, corresponds to the astral body of a human being? It is certainly not as easy to penetrate to the Earth's astral body as it is to its etheric body. We have seen that we can reach the etheric body if we allow the phenomena of the world to work upon us not merely through sense impressions but morally as well. If we wish to penetrate further, however, then deeper occult exercises are necessary, exercises such as you will find described in part—insofar as they can be in an open publication—in my book *How To Know Higher Worlds*. At a definite point of esoteric, or occult, development—as you can read there—a person begins to be conscious, even at a time when he or she is usually unconscious, namely, during the time between falling asleep and waking.

We know that the ordinary unconscious condition, the common human condition of sleep, is caused by our leaving our physical and etheric bodies lying in bed and drawing out our astral bodies and the rest of what belongs to them. The ordinary person is unconscious when this occurs. If, however, we devote ourselves more and more to those exercises that consist of meditation, concentration, and so on and further strengthen the slumbering hidden forces of our souls, we can establish a conscious condition of sleep. In that case, when we have drawn our astral bodies out of our physical and etheric bodies, we are no longer unconscious but have around us not the physical world, not even the world just described, namely, the world of the nature spirits, but another and still more spiritual world. When, after having freed ourselves from our physical and etheric bodies, the time comes that we feel our consciousness flash up, we then perceive quite a new order of spiritual beings.

The next thing that strikes the occult vision thus trained is that the new spirits we perceive have, as it were, command over the nature spirits. Let us be quite clear as to how far this is the case. I have told you that those beings whom we call the nature spirits of water work especially in the budding and sprouting plant world. Those that we may call the nature spirits of the air play their part in late summer and autumn, when the plants prepare to fade and die. Then the meteor-like air spirits sink down over the plant world and saturate themselves, as it were, with the plants, helping them to fade away in their spring and summer forms. The disposition, at one time, of the spirits of the water and, at another, the spirits of the air to work in this or that region of the Earth changes according to the different regions of the Earth—in the northern part of the Earth it is naturally quite different from what it is in the south. The task of directing the appropriate nature spirits to their proper activities at the right time is carried out by those spiritual beings whom we learn to know when occult vision is so far trained that when we have freed ourselves from our physical and etheric bodies, we can still be conscious of our environment.

There are spiritual beings, for instance, working in connection with our planet Earth, who allot the work of the nature spirits to the seasons of the year and thus bring about the alternations of these seasons for the different regions of the Earth, by distributing the work of the nature spirits. These spiritual beings represent what we may call the astral body of the Earth. We plunge into this astral Earth body with our own astral bodies at night when we fall asleep. This astral body, consisting of higher spirits that hover around the planet Earth and permeate it as a spiritual atmosphere, is united with the Earth, and into this spiritual atmosphere our astral bodies plunge during the night.

To occult observation, there is a great difference between the first category of nature spirits—the spirits of earth, water, air, and fire—and those beings that, on the other hand, direct these nature spirits. The nature spirits are occupied in causing the beings of nature to ripen and fade, in bringing life into the whole planetary sphere of the Earth. The situation is different with regard to those spiritual beings that in their totality can be called the astral body of the Earth. These beings are such that when we can become acquainted with them by means of occult vision, we perceive them as beings connected with our own souls—with our own astral bodies. They exert such an influence upon the human astral body—as also upon the astral bodies of animals—that we cannot speak of a mere life-giving activity; their activity resembles the action of feeling and thought upon our own souls. The nature spirits of water and air can be observed; we may say they are in the environment. But we cannot say of the spiritual beings of whom we are now speaking that they are in our environment; we are, in fact, always actually united with them, as if poured into them, when we perceive them. We are merged into them, and they speak to us in spirit. It is as though we perceived thoughts and feelings from the environment—impulses of will, sympathy and antipathy come to expression in what these beings cause to flow into us as thoughts, feelings, and impulses of will. Thus, in this category of spirits, we see beings already resembling the human soul.

If we turn back again to what has been stated earlier, we may say that all sorts of regulations in time—of divisions in the relations of time and space—are also connected with these beings. An old expression has therefore been preserved in occultism for these beings, which in their totality we recognize as the astral body of the Earth. In English this would be "spirits of the periods or cycles of time." Thus, not only the seasons of the year and the growing and the fading of the plants but also the regular alternation that in relation to the Planet Earth expresses itself in day and night are brought about by these spirits that are classed as belonging to the astral body of the Earth. In other words, everything connected with rhythmic return, with rhythmic alternation, with repetitions of happenings in time, is organized by spiritual beings who collectively belong to the astral body of the Earth and to whom the name "spirits of the periods or cycles of time of our planet" is applicable.

What the astronomer ascertains by calculating the cycle of the Earth on its axis is perceptible to occult vision, because the occultist knows that these spirits are distributed over the whole Earth and are actually the bearers of the forces that rotate the Earth on its axis. It is extremely important that one should be aware that in the astral body of the Earth is to be found everything connected with the ordinary alternations between the blossoming and withering of plants and also all that is connected with the alternations between day and night—between the various seasons of the year and the various times of the day and so on. In the observer who has progressed so far as to be able to leave the physical and etheric bodies in the astral body and still remain conscious, everything that happens in this way calls up an impression of the spiritual beings belonging to the spirits of the cycles of time.

We have now drawn aside the second veil, the veil woven of nature spirits. We might say that when we draw aside the first veil, woven of material physical impressions, we come to the etheric body of the Earth, to the nature spirits, and that then we draw aside a second veil and come to the spirits of the cycles of time, who regulate everything that is subject to rhythmical cycles. We know that what may be called the higher principle of human nature is embedded in our astral body. At first we understand this to be the "I," embedded in the astral body. We have already said that the astral body plunges into the region of the spirits of the cycles of time, that it is immersed in the surging sea of these spirits. But with respect to normal consciousness, the "I" is still more asleep than the astral body. One who is engaged in occult development and progressing esoterically becomes aware of this, because it is in the spiritual world into which one plunges— which consists of the spirits of the cycles of time—that one first learns to penetrate into the perceptions of the astral body.

In a certain respect, this perception is really a dangerous reef in esoteric development, for the human astral body is in itself a unity, but everything in the realm of the spirits of the cycles of time is, fundamentally, a multiplicity, a plurality. Because we are united with and immersed in this plurality in the way I have described, if we are still asleep in our "I" and yet awake in our astral body, we feel as if we were dismembered in the world of the spirits of the cycles of time. This must be avoided in a properly ordered esoteric development. Hence, those who are able to give instruction for such development see that the necessary precautions are taken so that one avoids, if possible, allowing the "I" to sleep when the astral body is already awake, for one would then lose inner cohesion and would, like Dionysius, be split up into the whole astral world of the Earth, consisting of the spirits of the cycles of time.

In a regular esoteric development, precautions are taken that this should not occur. These precautions consist in care being taken that the student, who through meditation, concentration, or other esoteric exercises is to be stimulated to clairvoyance, retains two things in the whole sphere of clairvoyant, occult observation. In every esoteric development it is especially important that everything should be adjusted such that two things we possess in ordinary life should not be lost—things we might, however, very easily lose in esoteric development if not rightly guided. If rightly guided, however, we will not lose them. First, we should not lose the recollection of any of the events of our present incarnation, as ordinarily retained in our memory. The connection with memory must not be destroyed. This connection with memory means very much more in the sphere of occultism than it does in the sphere of ordinary life. In ordinary life we understand memory only as the power of looking back and not losing consciousness of the important events of one's life. In occultism, a right memory means that we value with our perceptions and feelings only what we have already accomplished in the past, so that we apply no other value to ourselves or to our deeds than the past deeds themselves deserve.

Let us understand this quite correctly, for this is extremely important. If, in the course of one's occult development, one were suddenly driven to say to oneself, "I am the reincarnation of this or that spirit"—without there being any justification for it through any action of one's own—then one's memory, in an occult sense, would be interrupted. An important principle in occult development is that of attributing no other merit to oneself than what comes from one's actions in the physical world in the present incarnation. That is extremely important. Any other merit must come only on the basis of a higher development, which can be attained only if one first acknowledges firmly that one esteems oneself for nothing but what one has accomplished in this incarnation. This is quite natural if we look at the matter objectively, for what we have accomplished in the present incarnation is also the result of earlier incarnations; it is that which karma has so far made of us. What karma is still making of us we must first bring about; we must not add that to our value. In short, if we would set a right value on ourselves, we can do so at the beginning of esoteric development if we ascribe merit only to what is inscribed in our memory as our past. That is the one element we must preserve if our "I" is not to sleep while our astral body is awake.

The second thing that we as people of the present day must not lose is the degree of conscience we possess in the external world.

Here again is something that is extremely important to observe. You must have often experienced that someone you know has gone through an occult development and if it is not guided and conducted in the right way you find that, in relation to conscience, your friend takes things much more lightly than before the occult training. Before such training, education and social connection were the guides to whether one did this thing or that or dared not do it. After beginning an occult development, however, many people begin to tell lies who never did so before, and, with respect to questions of conscience, they take things more lightly. We ought not to lose an iota of the conscience we possess. In terms of memory, we must value ourselves only according to what we have already become, not according to any reliance on the future or on what we are still going to do. In terms of conscience, we must retain the same degree of conscience as we acquired in the ordinary physical world. If we retain these two elements in our consciousness—a healthy memory that does not deceive us into believing ourselves to be other than our actions prove us to be and a conscience that does not allow us morally to take things more lightly than before (indeed, if possible, we should take them more seriously!)—our "I" will never be asleep when our astral body is awake.

If we can remain awake in our sleep and preserve our consciousness and carry it with us into the condition wherein, with our astral body, we are freed from the physical and etheric bodies, then we shall carry the connection with our "I" into the world in which we awaken with our astral body. If we awake with our "I," not only do we feel our astral body to be connected with all the spiritual beings we have described as the spirits of the cycles of time of our planet, but we also feel in a quite peculiar way that we actually no longer have a direct relation to the individual who is the bearer of the physical and etheric bodies in which we usually live. We feel, so to speak, as if all the qualities of our physical and etheric bodies were taken from us. Then, too, we feel everything taken from us that can only live outwardly in a particular territory of our planet, for what lives in a particular region of our planet is connected with the spirits of the cycles of time. Now, however, when we waken with our "I," we feel ourselves not only poured out into the whole world of the spirits of the cycles of time but also one with the whole undivided spirit of the planet itself; we awaken in the undivided spirit of the planet itself.

It is extremely important that we should feel ourselves as belonging to the whole of our planet. For example, when our occult vision is sufficiently awakened and we are so far advanced that we can awaken our "I" and our astral body simultaneously, then our common life with the planet expresses itself in such a way that now, just as during our waking hours in the sense world we can follow the Sun as it passes over the heavens from morning till night, the Sun no longer disappears when we fall asleep. When we sleep, the Sun remains connected with us; it does not cease to shine but takes on a spiritual character, so that while we are actually asleep during the night, we can still follow the Sun. We human beings are of such a nature that we are connected with the changing conditions of the planet only insofar as we live in the astral body. When, however, we become conscious only of our "I," we have nothing to do with these changing conditions. But when we awaken our astral body and our "I" simultaneously, we become conscious of all the conditions that our planet can go through. We then pour ourselves into the whole substance of the planetary spirit.

When I say that we become one with the planetary spirit, that we live in union with this planetary spirit, you must not suppose that this implies an advanced degree of clairvoyance; this is but a beginning. For when we awaken in the manner described, we really experience the planetary spirit only as a whole, whereas it consists of many, many differentiations—of wonderful separate spiritual beings. The different parts of the planetary spirit, the special multiplicities of this spirit—of these we are not yet aware. What we realize first is the knowledge that can be expressed, "I live in the planetary spirit as though in a sea, which spiritually bathes the whole planet Earth and is the spirit of the whole Earth." One may go through immensely long development in order to experience further and further this unification with the planetary spirit, but to begin with the experience is as has been described.

Just as we say of ourselves as human beings, "Behind our astral body is our 'I'," so do we say that behind all that we call the totality of the spirits of the cycles of time is hidden the spirit of the planet itself, the planetary spirit. Whereas the spirits of the cycles of time guide the nature spirits of the elements in order to call forth the rhythmic changes and repetitions in time—the alterations in space of the planet Earth—the spirit of the Earth has a different task. It has the task of bringing the Earth itself into mutual relation with the other heavenly bodies in the environment, to direct it and guide it so that in the course of time it may come into the right relationship to the other heavenly bodies. The spirit of the Earth is the Earth's great sense apparatus, through which the planet Earth enters into proper relationship with the cosmos. If I were to sum up the succession of those spiritual beings with whom we on our Earth are first of all concerned and to whom we can find the way through a gradual occult development, I must say this: As the first external veil, we have the sense world, with all its multiplicity, with all we see spread out before our senses and which we can understand with our human mind. Then, behind this sense world, we have the world of nature spirits. Behind this world of nature spirits we have the spirits of the cycles of time, and behind these is the planetary spirit.

If we wish to compare what normal consciousness knows about the structure of the cosmos with the structure of the cosmos itself, we can make it clear in the following way. We will take it that the most external veil is this world of the senses, behind which is the world of the nature spirits and behind that the spirits of the cycles of time and behind that the planetary spirit. We must say that, in its activity, the planetary spirit in a certain respect penetrates through to the sense world, so that in a certain way we can perceive its image in the sense world. This also applies to the spirits of the cycles of time as well as to the nature spirits. If we observe the sense world itself with normal consciousness, we can see in the background, the impression, the traces, of those worlds that lie behind—as if we drew aside the sense world as the outermost skin and behind it we had different degrees of active spiritual beings. Normal consciousness realizes the sense world by means of its perceptions; the world of nature spirits expresses itself from behind these perceptions as what we call the forces of nature. When science speaks of the forces of nature, we have there nothing that is actually real; to the occultist the forces of nature are not realities but maya, imprints of the nature spirits working behind the world of sense.

Again, the imprint of the spirits of the cycles of time is what is usually known to ordinary consciousness as the laws of nature. Fundamentally, all the laws of nature are in existence because the spirits of the cycles of time work as the directing powers. To the occultist, the laws of nature are not realities. Whereas the ordinary natural scientist speaks of the laws of nature and combines them externally, the occultist knows that these laws are revealed in their reality when, in his awakened astral body, he listens to what the spirits of the cycles of time say and hears how they order and direct the nature spirits. That is expressed in maya—in external semblance—as the laws of nature, and normal consciousness as a rule does not go beyond this.

Normal consciousness does not usually reach the imprint of the planetary spirit in the external world. The normal consciousness of contemporary humanity speaks of the external world of perception, of the facts that can be perceived. It speaks of the forces of nature, light, warmth, magnetism, electricity, attraction and repulsion, gravity, and so forth. These are the beings of maya, behind which, in reality, lies the world of nature spirits—the etheric body of the Earth. External science also speaks of the laws of nature, but that again is maya. Underlying these laws is what we have described as the world of the spirits of the cycles of time. Only when we penetrate still further do we come to the stamp or imprint of the planetary spirit itself in the external sense world. Science today does not do this. Those who still do so are no longer quite believed. The poets, the artists do so—they seek for a meaning behind outer reality. Why does the plant world blossom? Why do the different species of animals arise and disappear? Why do human beings inhabit the Earth? If we thus inquire into the phenomena of nature and wish to analyze their meaning and to combine the external facts as even a deeper philosophy still sometimes tries to do, we approach the imprint of the planetary spirit itself in the external world. Today, however, nobody really believes any longer in this seeking after the meaning of existence. Through feeling, one still believes a little, but science no longer cares to know what could be discovered about the laws of nature by studying the passage of the phenomena.

If we still seek a meaning as to the laws of nature in the things of the world perceptible to our senses, we should be able to interpret this meaning as the imprint of the planetary spirit in the sense world, in external maya. The sense world itself is an external semblance, for it is what the etheric body of the Earth, the substance of the nature spirits, drives out of itself. A second maya is that which the nature spirits give shape to in the forces of nature. A third maya is what appears as the laws of nature, deriving from the spirits of the cycles of time. A fourth maya is something that, despite its maya nature, speaks to the human soul; in the perception of the purpose of nature, we feel ourselves united with the spirit of the entire planet, with the spirit that leads the planet through cosmic space and gives meaning to the whole planet. In this maya lies the direct imprint of the planetary spirit itself.

Thus we may say that we have today ascended to the undivided spirit of the planet. If we wish to compare what we have discovered about the planet with the human being, we may say this: "The sense world corresponds to the human physical body, the world of nature spirits to the human etheric body, the world of the spirits of the cycles of time to the human astral body, and the planetary spirit to the human 'I'." Just as the human "I" perceives the physical environment of Earth, so does the planetary spirit perceive everything in the periphery and in cosmic space as a whole outside the planet; it adjusts the acts of the planet and also the feelings of the planet according to these perceptions of cosmic space. For what a planet does in space when it passes on its way in cosmic distances and what it effects in its own body, in the elements of which it consists, are the result of the observations of the planetary spirit with respect to the external world. Just as the individual human soul lives in the world of the Earth side by side with other human beings and adjusts to them, so does the planetary spirit live in its planetary body—the ground on which we stand. This planetary spirit, however, lives in fellowship with other planetary spirits, other spirits of the heavenly bodies.

3

Elementals I

THE EVOLUTION OF PLANT LIFE, from the seed through budding and leafing, fruition, and withering away, is the work of a host of nature beings. Within the depths of the Earth, the gnomes encourage root growth, allowing for the sprouting of plants. As it buds and leafs and flowers and fruits, the plant takes into itself warmth and light and air from its surroundings and sends forces back down into the root. In a spiritual sense, the gnomes in their home underground receive these forces as secrets from the universe. These cosmic mysteries are no mysteries to the gnomes. The gnomes are perfectly attuned to them and understand them wholly. And while they are at home within the Earth, they harbor repugnance for what is earthly. They fear taking on earthly forms, in particular, the shape of amphibians. In their aversion, they drive plants upward, out of the ground that is the gnomes' home. There the undines, watery nature beings that live in the moist atmosphere, begin their work. These elementals, in their dreaming nature, assume an ever-changing variety of forms. They love the watery element, but as they flit around the plant, they try to avoid taking on the permanent shapes of fishes. In their own constant metamorphosis, they join and disperse substances around and within the plant and nurture its further development. As it continues to grow, the plant enters the airy, light-filled world of the sylphs, who feel a deep attraction to the birds that flit through the skies. In the vibrating currents of air produced by their flight, they hear the music of the spheres and absorb cosmic light, which they send down into the plants. Next, the growing plant enters the world of the fire spirits, once called salamanders. These spirits find an affinity with the butterflies and, indeed, all insects. They follow the flight of insects, as they flutter from plant to plant, carrying pollen, and they bathe the plant's flowers with cosmic warmth. Here we have a spiritual picture of the growth of plants. The gnomes push the plant upward from the womb of the Earth. The undines draw it out and weave chemical substances around it, while the sylphs imbue it with light; together they mold the plant's form—one might say the archetypal plant. When the plant dies and its physical matter falls away and enters the ground, the gnomes sense this ideal plant form that stands behind the physical substance; they protect it and prepare it for renewed life.

LECTURE BY RUDOLF STEINER[1]
NOVEMBER 2, 1923

To the outwardly perceptible, visible world there belongs the invisible world, and these worlds, taken together, form a whole. The marked degree to which this is the case first appears in its full clarity when we turn our attention away from the animals to the plants. Plant life, as it sprouts and springs forth from the Earth, immediately arouses our delight, but it also provides access to something that we feel is full of mystery. Although the will and

---

1. Lecture 7 from *Harmony of the Creative Word: The Human Being and the Elemental, Animal, Plant, and Mineral Kingdoms*, twelve lectures, Dornach, Oct. 19–Nov. 11, 1923 (GA 230).

whole inner activity of the animal certainly contain something of the mysterious, we nevertheless recognize the presence of this will and that it is the origin of the animal's form and outer characteristics. In the case of plants, which appear on the face of the Earth in such magnificent variety of form and which develop in such a mysterious way out of the seed, with the help of earth and the surrounding air, we feel that some other factor must be at work that this plant world may arise in the form it does.

When spiritual vision is directed to the plant world, we are immediately led to a whole host of beings that were known and recognized in the old times of instinctive clairvoyance but which were afterward forgotten and today remain only as names used by poets, names to which modern human beings ascribe no reality. To the same degree, however, to which we deny reality to the beings that flit so busily around the plants, to that degree do we lose understanding of the plant world. This understanding of the plant world, which, for instance, is so necessary to the practice of medicine, has been entirely lost to present-day humanity.

We have already recognized a very significant connection between the world of plants and the world of butterflies, but this understanding, too, will really come alive for us only when we look yet more deeply into the whole range of activities and processes that go on in the plant world. Plants send down their roots into the ground. Anyone who can observe what they send down and can perceive the roots with spiritual vision (which is essential) sees how the root is everywhere surrounded by the activities of elemental nature spirits. These elemental spirits, which an old clairvoyant perception designated as gnomes and which we may call the root spirits, can actually be studied through Imagination and Inspiration, just as human life and animal life can be studied in the physical world. We can look into the soul nature of these elemental spirits, into this world of the spirits of the roots.

The root spirits are quite special earth folk, invisible at first to outer view but in their effects so much the more visible. No root could develop if it were not for what is mediated between the root and the earth realm by these remarkable root spirits, which bring the mineral element of the Earth into flux in order to conduct it to the roots of plants. I am, of course, referring to the underlying spiritual process. These root spirits, which are everywhere present in the Earth, gain a quite particular sense of well-being from rocks and ores (which may be more or less transparent and also contain metallic elements). They have the greatest feeling of well-being in this sphere, because it is the place where they belong, where they are conveying what is mineral to the roots of the plants. They are filled with an inner spirituality that we can compare only to the inner spirituality of the human eye and the human ear. For these root spirits are in their spiritual nature entirely sense. Apart from this, they are nothing at all; they consist only of sense. They are completely sense, and it is a sense that is at the same time *intellect*, which does not simply see and hear but also immediately understands what is seen and heard; it not only receives impressions but everywhere also receives ideas.

We can even indicate the way in which these root spirits receive their ideas. We see a plant sprouting out of the Earth. The plant enters, as I shall presently show, into connection with the extraterrestrial universe. Spiritual currents flow from above, particularly at certain seasons of the year, from the flower and the fruit of the plant down into the root, streaming into the Earth. And just as we turn our eyes toward the light and see, so do the root spirits turn their faculty of perception toward what trickles downward from above, through the plant into the Earth. What trickles down toward the root spirits is something that the light has sent into the flowers, which the heat of the Sun has sent into the plants, which the air has produced in the leaves, and which the distant stars have brought about in creating the plant form. The plant gathers the secrets of the universe and sends them into the ground, and the gnomes take these secrets into themselves from that which trickles down spiritually to them through the plants. Because the gnomes in their wanderings through ore and rock bear with them what has come down to them through the plants, particularly from autumn through the winter, they are the beings within the Earth that carry the ideas of the whole universe on their streaming, wandering journey through the Earth.

We look out into the wide world. The world has been built of the spirit of the universe; it embodies the ideas of the universe. Through the plants, which to the gnomes are the same as rays of light are to us, the gnomes take in the ideas of the universe and carry them in full consciousness from metal to metal, from rock to rock within the Earth. We gaze down into the depths of the Earth not to seek there for abstract ideas about mechanical laws of nature but to behold the roving, wandering gnomes, which are the light-filled preservers of cosmic reason within the Earth. These gnomes have immediate understanding of what they see; their knowledge is of a similar nature to that of human beings, but they are the epitome of intellect; they are nothing but intellect. Everything about them is intellect, but an intellect so universal that they look down on human reason as something imperfect. The gnomes laugh us to scorn on account of the groping, struggling intellect with which we manage to grasp one idea or another, whereas they have no need at all to think things through. They have direct perception of what is sensible and intelligent in the world, and they are particularly ironic when they notice the efforts people have to make to come to this or that conclusion. Why should they do this? say the gnomes; why should people give themselves so much trouble to think things over? We know everything we look at. People are so stupid, say the gnomes, for they must first think matters over.

Gnomes can be ironic to the point of ill manners if one speaks to them of logic. For why should people need such a superfluous thing—a training in thinking? The thoughts are already there. The ideas flow through the plants. Why do not people stick their noses into the Earth down to the depth of the plant's roots and let what the Sun says to the plants filter into their noses? Then they would know something! But with logic, say the gnomes, one can have only odd bits and pieces of knowledge. Thus the gnomes are actually the bearers of the ideas of the universe, of the cosmos, inside the Earth, but for the Earth itself they have no liking at all. They flit about in the Earth with cosmic ideas, but they actually hate what is earthly. This is something from which the gnomes would best like to escape. Nevertheless they remain with the earthly—we will soon see why this is so—but they hate it, for the earthly threatens them with continual danger.

The Earth continually holds over them the threat of forcing them to take on particular shapes, the configuration of the creatures I described in the last lecture, the amphibians and, in particular, frogs and toads. The feeling of the gnomes within the Earth is this: "If we grow too strongly together with the Earth, we shall assume the form of frogs or toads." They are continually on the alert to avoid being caught up too strongly in the Earth and being forced to take on such an earthly form. They are always on the defensive against this earthly form, which threatens them in the element in which they exist. They have their home in the element of earth and moisture; there they live under the constant threat of being forced into amphibian forms. From this they repeatedly tear themselves free by filling themselves entirely with ideas of the extraterrestrial universe. The gnomes are the element within the Earth that represents the extraterrestrial, because they must always avoid growing together with the earthly; otherwise they would individually take on the forms of the amphibian world. And it is just from what I may call this feeling of hatred, this antipathy toward the earthly, that the gnomes gain the power of driving the plants up from the Earth. With the fundamental force of their being they unceasingly thrust away from the earthly, and it is this thrust that determines the upward direction of plant growth; they impel the plants along with them.

The antipathy that the gnomes have to anything earthly causes the plant to have only its roots in the Earth and then grow out of the Earth; in fact, the gnomes force the plants out of their true, original form and make them grow upward and out of the Earth.

Once the plant has grown upward, once it has left the domain of the gnomes and has passed out of the sphere of the element of moist earth into the sphere of moist air, the plant develops what comes to outer physical form in the leaves. Other beings then go to work in everything that goes on in the leaves—water spirits, elemental spirits of the watery element, to which an earlier instinctive clairvoyance gave the name of undines, among others. Just as we found gnomes flitting busily around the roots, we find close to the soil elemental beings of water, the undines, who observe with pleasure the upward-striving growth that the gnomes have produced.

These undines differ in nature from the gnomes. They cannot turn outward toward the universe like a spiritual sense organ. They can only yield themselves up to the movement and activity of the whole cosmos in the element of air and moisture, and they therefore do not have the clarity of mind that the gnomes have. They dream incessantly, these undines, but their dream is at the same time their own form. They do not hate the Earth as intensely as the gnomes, but they have a sensitivity to what is earthly. They live in the etheric element of water, swimming and floating in it. They are highly sensitive to anything in the nature of a fish, for the fish's form is a threat to them. They do assume this form from time to time, though only to forsake it immediately to take on another metamorphosis. They dream their own existence, and in dreaming their own existence they bind and release, unite and separate, the substances of the air, which in a mysterious way they introduce into the leaves. They take these substances to the plants that the gnomes have thrust upward.

The plants would wither at this point if it were not for the undines, who approach from all sides. As they move around the plants in their dreamlike consciousness, they prove to be what we can only call world chemists. The undines dream the binding and releasing of substances. This undine dream, through which the plant exists and into which it grows when, developing upward, it leaves the ground, is the work of the world chemist inducing the mysterious combining and separating of substances in the plant world, starting in the leaf. We can therefore say that the undines are the chemists of plant life. They dream of chemistry. They possess an exceptionally delicate spirituality that is really in its element just the place where water and air come into contact with each other.

The undines live entirely in the element of moisture, but they have a sense of inner satisfaction when they approach the surface of something watery, be it only a drop of water or something else of a watery nature. For their whole endeavour lies in preserving themselves from totally assuming the form of a fish, the permanent shape of a fish. They wish to remain in a state of transformation, of eternal, endlessly continuing changeability. In this state of changeability, in which they dream of the stars and of the Sun, of light and of heat, they become the chemists who, starting from the leaf, carry on the further development of the plant form that has been thrust upward by the gnomes. In this way the plant develops leaf growth, and this mystery is revealed as the dream of the undines into which the plants grow.

To the same degree, however, in which the plant grows into the dream of the undines, it now enters into another, higher domain, the domain of the spirits that live in the element of air and warmth—just as the gnomes live in that of earth and moisture and the undines in that of air and moisture. In the element of air and warmth live the beings that an earlier clairvoyant faculty called the sylphs. Because air is everywhere imbued with light, these sylphs living in the element of air and warmth press toward the light and become related to it. They are particularly susceptible to more delicate but more widespread movements within the atmosphere.

When in spring or autumn one sees a flock of swallows, which produce vibrations in a body of air as they fly along, creating a current of air, this moving air current—and this holds true for every bird—is something the sylphs can hear. Within the current of air they hear cosmic music. If, let us say, one is traveling somewhere by ship and the seagulls fly toward it, their flight sets in motion spiritual sounds, a spiritual music that accompanies the ship.

It is the sylphs that unfold and develop their being within this sounding music, finding themselves at home in the moving current of air. Within this spiritually sounding, moving element of air they find themselves at home; at the same time they absorb what the power of light sends into these vibrations of the air. The sylphs, which experience existence more or less in a state of sleep, feel most in their element, most at home, where birds are winging through the air. If a sylph is obliged to flit through air devoid of birds, it feels as though it has lost itself. But at the sight of a bird in the air, a quite special feeling comes over the sylph. I have often had to describe a certain event in human life, the event that leads the human soul to address itself as "I." I have always drawn attention to the statement made by Jean Paul: that, when a human being first arrives at the conception of the "I," it is as though he or she looks into the most deeply veiled Holy of Holies of his soul. A sylph does not look into any such veiled Holy of Holies of its own soul, but when it sees a bird, a feeling of ego comes over it. The sylph feels its ego through what the bird sets in motion as it flies through the air. And because this is so, because its ego is kindled in it from outside, the sylph becomes the bearer of cosmic love through the atmosphere. It is because the sylph embodies something like a human wish, but does not have its ego within itself but in the bird kingdom, that it is at the same time the bearer of wishes of love through the universe.

Thus we behold the deepest sympathy between the sylphs and the bird world. The gnome hates the amphibian world; the undine is sensitive to fishes, is unwilling to approach them, seeks to avoid them and feels a kind of horror for them. The sylph, on the other hand, is attracted to birds and has a sense of well being when it can waft toward their feathered flight the floating air filled with sound. Were one to ask a bird from whom it learns to sing, one would hear that its inspirer is the sylph. Sylphs feel a sense of pleasure in the bird's form. They are, however, prevented by the cosmic order from becoming birds, for they have another task. Their task is lovingly to convey light to the plant. Just as the undine is the chemist for the plant, so is the sylph the light bearer. The sylph imbues the plant with light; it bears light into the plant.

Through the fact that the sylphs bear light into the plant, something quite remarkable is brought about. You see, the sylph is continually carrying light into the plant. The light, that is to say the power of the sylphs in the plant, works on the chemical forces that were induced in the plant by the undines. Here takes place the interworking of the sylph's light and the undine's chemistry. This is a remarkable moulding and shaping activity. With the help of the upstreaming substances that are worked on by the undines, the sylphs weave an ideal plant form out of the light. They actually weave the archetypal plant within the plant from light and from the chemical working of the undines. When toward autumn the plant withers and everything of physical substance disperses, then these forms of the plants begin to trickle downward, and the gnomes perceive them; they perceive what the world—the sun through the sylphs, the air through the undines—has brought to pass in the plant. This the gnomes recognize; throughout the entire winter they are engaged in perceiving below what has trickled down into the soil from the plants. Down there they grasp world ideas in the plant forms that have been given shape and form with the help of the sylphs and which now enter into the soil in their spiritual, ideal form.

People who regard the plant as something material will, of course, know nothing of this spiritual ideal form. Thus, at this point, a colossal error, a terrible error appears in materialistic observation of the plant. I'll give a brief outline of it. Everywhere one will find that in materialistic science matters are described as follows: The plant takes root in the ground, above the ground it develops its leaves and finally its flowers, and within the flower the stamens, then the carpel. The pollen from the anthers, usually from another plant, is taken over to the stigma; the carpel is fertilized, and through this the seed of the new plant is produced. That is the usual way of describing the process. The carpel is regarded as the female element and what comes from the stamens as the male. Indeed, matters cannot be regarded otherwise as long as people remain bound to materialism, for then this process really does look like fertilization. This, however, it is not. To gain insight into the process of fertilization, that is, the process of reproduction, in the plant world, we must be conscious that in the first place the plant form arises through the work of the undines and the work of the sylphs. The ideal plant form goes down into the ground and is kept safe by the gnomes. It is there below ground, this plant form. Within the Earth it is guarded by the gnomes after they have seen and perceived it. The Earth becomes the womb for what thus trickles downward. This is something quite different from what is described in materialistic science.

After it has passed through the sphere of the sylphs, the plant enters the sphere of elemental fire spirits. These spirits inhabit the element of heat and light. When the warmth of the Earth is at its height, or has reached a sufficient level, it is gathered up by the fire spirits. Just as the sylphs gather up the light, so do the fire spirits gather up the warmth and carry it into the flowers of the plant.

Undines carry the action of chemical ether into the plants, sylphs the action of light ether into the flowers. And the pollen provides what may be called little airships that enable the fire spirits to carry warmth into the seed. Everywhere warmth is collected with the help of the stamens and is carried by means of the pollen from the anthers to the seeds in the carpel. And what is formed here in the carpel in its entirety is the male element that comes from the cosmos. It is not a case of the carpel being female and the anthers of the stamens being male. In no way does fertilization occur in the flower, but only the preforming of the male seed. Fertilization takes place when the cosmic male seed, which fire spirits in the flower take from the warmth of the universe, is brought together with the female principle that has trickled down into the soil as an ideal element, at an earlier stage, and is resting there.

For plants the Earth is the mother and the heavens the father. And all that takes place outside the domain of the Earth is not the maternal womb for the plant. It is a colossal error to believe that the maternal principle of the plant is in the carpel. This is, in fact, the male principle, which has been drawn forth from the universe with the aid of the fire spirits. The maternal element is taken from the cambium of the plant, which lies between bark and wood and is carried down as ideal form. What results from the combined activity of the gnomes and fire spirits is fertilization. The gnomes are, in fact, the spiritual midwives of plant reproduction. Fertilization takes place within the Earth during the winter, when the seed enters the Earth and meets with the forms that the gnomes have received from the activities of the sylphs and undines and which they then carry to where these forms can meet with the fertilizing seeds.

Because people do not recognize what is spiritual and do not know that gnomes, undines, sylphs, and fire spirits (which were formerly called salamanders) are actively involved in plant growth, there is a complete lack of clarity about the process of fertilization in the plant world. Above ground nothing by way of fertilization takes place; the Earth is the mother of the plant world and the heavens are the father. This is the case in a quite literal sense. Plant fertilization takes place through the fact that gnomes take from fire spirits what the fire spirits have carried into the carpel as concentrated cosmic warmth on the tiny airships of the anther pollen. Thus, the fire spirits are the bearers of warmth.

Now we will easily gain insight into the whole process of plant growth. First, with the help of what comes from the fire spirits, the gnomes down below instill life into the plant and push it upward. They are the fosterers of life. They carry the life ether to the root, the same life ether in which they themselves live. The undines foster the chemical ether in the plant, the sylphs the light ether, and the fire spirits the warmth ether. Then the fruit of the warmth ether unites with what is present below the ground as life. Plants can be understood only when they are considered in connection with all that is flitting around them full of life and activity. One reaches the right interpretation of the most important process in the plant only when one penetrates into these matters in a spiritual way.

Once this has been understood, it is interesting to look again at the words Goethe jotted down when responding to another botanist. He was terribly annoyed because people talked about endless "marriages" going on in the plants. Goethe was affronted by the idea that marriages were constantly being consummated all over every meadow. This seemed to him unnatural. In this Goethe had an instinctive, but very true feeling. He could not as yet know the real facts of the matter, but nevertheless his instinct was a sure one. He could not see why fertilization should take place in the flower. He did not as yet know what goes on below ground and that the Earth is the maternal womb of the plants, but he intuitively knew that the process that takes place in the flower is not what all botanists take it to be. You are now aware of the inner connection between plant and Earth, but here is something else which must be taken into account.

When up above the fire spirits are flitting around the plant and transmitting the pollen from the anthers, they have only one feeling; they have this feeling to an enhanced degree, compared with the feeling of the sylphs. The sylphs experience their self, their ego, when they see the birds flit about. The fire spirits have a similar, but intensified experience with respect to the butterfly world and, indeed, the insect world as a whole. It is these fire spirits that take the utmost delight in following in the tracks of the insects' flight, so that they convey warmth to the carpel. To carry the concentrated warmth, which must descend into the Earth so that it may be united with the plant's ideal form, the fire spirits feel themselves intimately related to the butterfly world and to the world of the insects in general. Everywhere they follow in the tracks of the insects as they flit from flower to flower. One has the feeling, when following the flight of insects, that each of these insects, as it flutters from flower to flower, has a quite special aura that cannot be entirely explained in terms of the insect itself. The luminous, wonderfully radiant, shimmering aura of bees, as they dart from flower to flower, is unusually difficult to explain. Why? It is because the bee is everywhere accompanied by a fire spirit that feels so closely related to it that, seen with spiritual vision, the bee is surrounded by an aura that is in fact a fire spirit. When a bee flies through the air from plant to plant, from tree to tree, it flies with an aura that is imparted to it by a fire spirit. The fire spirit does not simply gain a feeling of its ego in the presence of the insect but actually wishes to be completely united with the insect.

Through this connection, insects also obtain that power about which I have spoken, which even shows itself in a shimmering forth of light into the cosmos. They obtain the power to spiritualize completely the physical matter that unites itself with them and to allow this spiritualized physical substance to ray out into cosmic space. Just as heat is what first causes light to shine, so, above the surface of the Earth, it is the fire spirits that flit around them that inspire the insects to bring about this shimmering forth into cosmic space—which attracts the human being to descend again into physical incarnation. If the fire spirits are active in promoting the outstreaming of spiritualized matter into the cosmos, they are no less actively engaged in seeing to it that the concentrated fiery element, concentrated warmth, goes into the interior of the Earth, so that with the help of the gnomes, the spirit form—which sylphs and undines caused to trickle down into the Earth—may be awakened.

This is the spiritual process of plant growth. It is because the subconscious in human beings divine something of a special nature in the flowering, sprouting plant that they experience the being of the plant as full of mystery. This is not to dismantle the mystery, of course, nor brush the dust from the butterfly's wings. Our instinctive delight in the plant is raised to a higher level when not only the physical plant is seen but also that wonderful working of the gnomes' world below, with its immediate understanding that gives rise to intelligence, the gnomes' world that first pushes the plant upward. Just as the human intellect is not subject to gravity and the head is carried without our feeling its weight, so the gnomes, with the light of their intellect, overcome what is of the Earth and push the plant upward. They prepare life down below, but the life would die away were it not given impetus by the chemical activity brought to it by the undines. This activity must be imbued with light.

We picture coming upward from below, in bluish, blackish shades, the force of gravity, to which an upward impulse is given by the gnomes. Flitting all around the plant—as indicated by its leaves—the undine power blends and disperses substances as the plant grows upward. From above downward, through the work of the sylphs, light is made to leave its imprint in the plant and moulds and creates the form that descends as an ideal form and is taken up by the maternal womb of the Earth. Moreover, fire spirits flit around the plant and concentrate cosmic warmth in the tiny seed points. This warmth is sent down to the gnomes together with the seed power, so that below ground they can cause the plants to arise out of fire and life.

We now see that the Earth is indebted for its power of repulse and impulse for growth and its density to the antipathy of the gnomes and undines toward amphibians and fishes. If the Earth is dense, this density is due to the antipathy by means of which the gnomes and undines maintain their forms. When light and warmth come down to Earth, this is an expression of that power of sympathy, that sustaining power of sylph love, carried through the air and to the sustaining sacrificial power of the fire spirits, which brings the power to descend to what lies below ground. We may say that over the face of the Earth, density, magnetism and gravity, in their upwardly striving aspect, unite with the downward striving power of love and sacrifice. In this interworking of the downward-streaming force of love and sacrifice and the upward-streaming force of density, gravity, and magnetism, where the two streams meet, plant life develops on the surface of the Earth. Plant life is an outer expression of the interworking of world love, world sacrifice, world gravity and world magnetism.

Now we have seen what matters when we direct our gaze to the plant world, which so enchants, uplifts, and inspires us. Real insight can be gained only when our vision embraces the spiritual, the supersensible, as well as that which is accessible to the physical senses. This enables us to correct the capital error of materialistic botany, that fertilization takes place above the Earth. What occurs there is not the process of fertilization but the preparation of the plant's heaven-born male seed for the future plant being nurtured in the maternal womb of the Earth.

4

Elementals II

ELEMENTAL NATURE BEINGS have certain relationships to the animal world as well as to the plant world. The gnomes serve as a complement to amphibians, animals that have no internal supporting structure and live in a minimal state of consciousness. The gnomes, by contrast, are wide awake to their world and take on and constantly renew their own form. They spiritually "support" the structure of amphibians, in a manner of speaking. So, too, the undines relate to fishes. The undines' activity gives them their external structure of scales and shells. Birds, which have only stunted legs, have their complement in the sylphs who have complete spiritually developed "limbs." The counterpart of the fire spirits we find in the butterflies, in their insubstantiality.

Ordinarily, we are not awake to the world of elementals. In our ordinary consciousness, the limitations of our physical bodies prevent us from interacting with the nature spirits. However, although we are not fully aware of it, we can begin to perceive these beings when we sleep. We find the gnomes in the pictures we form as we fall asleep. The undines rise to meet us from the sea of dreamless deep sleep. The sylphs greet us in the images that arise when we verge on awakening. We meet with the fire spirits in waking consciousness, when we are able to step back from ourselves as thinkers and see these spirits at work revealing our thoughts to us, bringing them to us from the cosmic spaces.

But the world of the elementals contains harmful as well as beneficent spirits. The malevolent nature spirits enter realms where they should not be active, bringing down to Earth forces that properly belong in the heavens or sending up to the heavens forces that belong to Earth. These forces of dissolution and rejuvenation, when transposed to realms to which they do not belong, have negative effects both within the human being and in the world around us. Only when we gain the proper insight can we distinguish between these beneficent and malevolent beings and recognize their working and interweaving in the world.

LECTURE BY RUDOLF STEINER[1]
NOVEMBER 3, 1923

Yesterday I spoke about the other side of the natural world, about the supersensible and invisible beings that accompany the beings and processes visible to the senses. An earlier, instinctive vision beheld the beings of the supersensible world as well as those in the world of the senses. Today, these beings have withdrawn from human view. The reason why this company of gnomes, undines, sylphs, and fire spirits is not perceptible in the same way as animals, plants, and so on is merely that the human being, in the present epoch of Earth evolution, is not in a position to unfold the soul and spirit without the help of the physical and ether bodies. In the present situation of Earth evolution humans are obliged to depend on the etheric body for the purposes of the

1. Lecture 8 from *Harmony of the Creative Word: The Human Being and the Elemental, Animal, Plant, and Mineral Kingdoms.*

soul and on the physical body for the purposes of the spirit. The physical body, which provides the instrument for the spirit, that is, the sensory apparatus, is not able to enter into communication with the beings that exist behind the physical world. It is the same with the etheric body, which humans need to develop as an ensouled being. Thus, half of earthly environment escapes us. We pass over everything connected with the elemental beings about which I spoke yesterday. The physical and the ether body have no access to this world. We can gain an idea of what actually escapes human beings today when we realize the nature of gnomes, undines, sylphs, and salamanders.

We have a whole host of lower animals—lower at the present time. These animals consist only of a soft mass, live in the fluid element, and have nothing in the way of a skeleton to give them internal support. They are creatures that belong to the latest phase of the Earth's development; creatures that only now, when the Earth has already evolved, develop what the human being, the oldest Earth being, already developed in his head structure during the time of ancient Saturn. These creatures have not progressed to the stage of producing the hardened substance in them that can become supporting skeleton.

It is the gnomes that, in a spiritual way, make up in the world for what is lacking in the lower orders of the animals up to the amphibians and fish, which have only the beginnings of a skeleton. These lower animal orders become complete, as it were, only through the fact that gnomes exist. The relationships between beings in the world are very different, and something arises between these lower creatures and the gnomes that I yesterday called antipathy. The gnomes do not wish to become like these lower creatures. They are continually on the watch to protect themselves from assuming their form. The gnomes are extraordinarily clever, intelligent beings. With them intelligence is already implicit in perception; they are in every respect the antithesis of the lower animal world. They have a certain significance for plant growth, but in the case of the lower animal world they actually provide its complement. They supply what this lower animal world does not possess.

This lower animal world has a dull consciousness; the gnomes have a consciousness of the utmost clarity. The lower creatures have no bony skeleton, no bony support; the gnomes bind together everything that exists by way of gravity and fashion their bodies from this volatile, invisible force, bodies that are, moreover, in constant danger of disintegrating, of losing their substance. The gnomes must ever and again create themselves anew out of gravity, because they continually stand in danger of losing their substance. For this reason, in order to preserve their own existence, the gnomes are constantly attentive to what is going on around them. No being is a more attentive observer on Earth than a gnome. It takes note of everything, for it must know everything, grasp everything, in order to survive. A gnome must always be wide awake; if it were to become sleepy, as people often do, this sleepiness would immediately cause its death.

There is a German saying of very early origin that aptly expresses this characteristic of the gnomes, in having always to remain attentive. People say: "Look sharp like a goblin." The gnomes are goblins. If one wishes to make someone attentive, then one says to him: "Look sharp like a gnome." A gnome is a truly attentive being. If one could place a gnome as an object lesson on a front desk in every school classroom, where all could see it, it would be a splendid example for the children to follow.

The gnomes have yet another characteristic. They are filled with an absolutely unconquerable lust for independence. They trouble themselves little about one another and give their attention only to the world of their own surroundings. One gnome takes little interest in another, but everything else in this world around them, in which they live, this interests them exceedingly. The human body is a hindrance to our perceiving such folk as the gnomes. The moment it ceases to be this hindrance, these beings are visible, just as the other beings of nature can be seen with ordinary vision. Anyone who reaches the stage of experiencing dreams in full consciousness on falling asleep is well acquainted with the gnomes. You need only recall what I recently published in the *Goetheanum* on the subject of dreams.[1] I said that a dream in no way appears to ordinary consciousness in its true form, but wears a mask. Such a mask is worn by the dream that we have on falling asleep. We do not immediately escape from the experience of our ordinary day consciousness. Reminiscences well up, memory pictures from life. Or we perceive symbols representing the internal organs—the heart as a stove, the lungs as wings. These are masks. If we were to see a dream unmasked, if we were actually to pass into the world of sleep without the beings existing there being masked, then, at the moment of falling asleep, we would behold a whole host of goblins coming toward us.

In ordinary consciousness human beings are protected from seeing these things unprepared, for they would terrify them. The form in which these beings would appear would actually be reflections, images of all the qualities in a particular individual that work as forces of destruction. A person would perceive all the destructive forces within himself or herself, all that continually destroys. These gnomes, if perceived by someone who is unprepared, would be nothing but symbols of death. Humans would be terribly alarmed by them if in ordinary consciousness they knew nothing about them and were confronted by them on falling asleep. They would feel entombed by them, yonder in the astral world. Seen from the other side, what takes place on falling asleep is a kind of entombment by gnomes.

This holds good only for the moment of falling asleep. A further complement to the physical, sense perceptible world is given by the undines, the water beings, which continually transform themselves and which live in connection with water just as the gnomes live in connection with the Earth. These undines, whose role in plant growth we have come to know—exist as complementary beings to

2. Rudolf Steiner. *On the Life of the Soul*, Anthroposophic Press, N.Y., 1985.

animals that are at a somewhat higher stage and have assumed a more differentiated earthly body. These animals, which have developed into the more evolved fishes or into the more evolved amphibians, require scales, that is, some sort of hard external armour. For the powers needed to provide certain creatures with this outer support, or external skeleton, the world is indebted to the activity of the undines. The gnomes spiritually support the creatures that are at quite a low evolutionary stage. The creatures that must be supported externally, must be clad in a kind of armour, owe their protective armour to the activity of the undines. Thus it is the undines that impart to these somewhat higher animals, in a primitive way, what we have in the cranial portion of the skull. They make them, as it were, into heads.

All the beings that are invisibly present behind the visible world have their vital task in the great scheme of things. One will always notice that materialistic science breaks down when one tries to use it to explain something of the kind I have just described. It cannot, for instance, explain how the lower creatures, which are scarcely any more solid than the element in which they live, manage to propel themselves forward in it, because scientists do not know about the spiritual support provided by the gnomes. Equally, the development of an external shell will always present a problem to purely materialistic scientists, because they do not know that the undines, in their sensitivity to and avoidance of their own tendency to become lower animals, cast off what then appears on the somewhat higher animals as scales or some other kind of outer shell.

In the case of these elemental beings, it is only our body that hinders the ordinary consciousness of today from seeing them, just as it sees the leaves of plants or the somewhat higher animals. When, however, we fall into a state of deep, dreamless sleep—a sleep that is, in fact, not dreamless, because through the gift of Inspiration it has become transparent—our spiritual gaze perceives the undines rising up out of that astral sea in which, on falling asleep, we were engulfed, submerged by the gnomes. In deep sleep the undines become visible. Sleep extinguishes ordinary consciousness, but the sleep that is illumined by clear consciousness is filled with the wonderful world of ever-changing fluidity, a fluidity that rises up in all kinds of ways to create the metamorphoses of the undines. Just as our day consciousness perceives creatures around us that have firm contours, a clear night consciousness would present to us these ever-transforming beings, which themselves rise up and sink down again like the waves of the sea. All deep sleep is filled with a moving sea of living beings, a shifting sea of undines all around the human being.

It is different in the case of the sylphs. In a way they also complement the being of certain animals, but in the other direction. The gnomes and undines add what is of the nature of the head to the animals, where it is lacking. Birds, however, are pure head; they are entirely head. The sylphs add to the birds in a spiritual way what they lack as the physical complement of the head. They balance the bird kingdom with what corresponds to the metabolism and limbs in humans. If the birds fly about in the air with atrophied legs, so much the more powerfully developed is the limb system of the sylphs. They may be said to represent in the air, in a spiritual way, what the cow represents below in physical matter. This is why I was able to say yesterday that it is in connection with the birds that the sylphs have their ego, that they have what connects them with the Earth. Human beings acquire an ego on the Earth. The bird kingdom connects the sylphs with the Earth. The sylphs are indebted to the bird kingdom for their ego, or at least for the consciousness of their ego.

Let us suppose we have slept through the night, that we have had around us the astral sea that gives rise to the most manifold undine forms, and then perceive a dream on awakening. If this dream we wake up to were not masked in things reminiscent of life or in symbolic representations of the human organs, if we were to see the unmasked dream, we would be confronted by the world of the sylphs. But these sylphs would assume for us a strange form; they would appear much as the Sun might appear if it wished to impart to human beings something that would affect them strangely, something that would have a soporific spiritual effect on their sleep. We shall hear in a moment why this is so. If we were to perceive an unmasked dream on awakening, we would see light fluttering toward us, the essence of light. We would find this an unpleasant experience, particularly because the limbs of these sylphs would, as it were, spin and weave around us. We would feel as though the light were attacking us from all sides, as if the light were something that beset us and to which we were extraordinarily sensitive. Now and then we would perhaps also feel the light to be like a caress. What I really wish to indicate here is only that the light, with its uplifting, gently stroking quality, approaches in the sylph's form.

When we come to the fire spirits, we find that they complement the fleeting nature of the butterflies. A butterfly itself develops as little as possible of its actual physical body; it leaves the body as tenuous as possible. It is a creature of light. The fire spirits appear as beings that complement the butterfly's body, so that we can gain the following impression. Suppose we have, on the one hand, a physical butterfly before us and pictured it suitably enlarged and, on the other hand, a fire spirit—they are, it is true, rarely together, except in the circumstances that I mentioned yesterday. If these two beings were welded together, we would have something resembling a winged human being. We need only increase the size of the butterfly and adapt the size of the fire spirit to human proportions, and from this we would get something like a winged human being.

The fire spirits are the complement to creatures nearest to what is spiritual; they complement them in a downward direction. Gnomes and undines complement in an upward direction, toward the head; sylphs and fire spirits complement the birds and butterflies in a downward direction. Thus the fire spirits must be seen in connection with the butterflies. In the same way that human beings can penetrate the dreams of sleep, so can they also penetrate waking day life. But there they make use of the physical body in quite a robust way. This, too, I have described in essays in the *Goetheanum* mentioned previously.[2] We do not realize that we could actually see the fire spirits in our waking life, for the fire spirits are inwardly related to our thoughts, to everything that proceeds from the head. When we have progressed so far that we can remain in waking consciousness but nevertheless stand in a certain sense outside the self, viewing it from outside as a thinking being while standing firmly on the Earth, then we will become aware of how the fire spirits are the element in the world that makes our thoughts perceptible to us from the other side.

Thus the perceiving of the fire spirits enables us to see ourselves as thinkers, not merely to be the thinker and, as such, hatch our thoughts. We can actually behold how the thoughts run their course. In that case, however, thoughts cease to be bound to the human being. They reveal themselves to be world thoughts; they are active and move as impulses in the world. Then one finds that the human head calls forth only the illusion that thoughts are enclosed inside the skull. They are simply reflected there; their mirrored images are there. What underlies these thoughts belongs to the sphere of the fire spirits. On entering this sphere, one sees thoughts to be not only what they are in themselves but also the thought content of the world, which is rich in imaginative content. The power to stand outside oneself is the power that enables one to arrive at the realization that thoughts are world thoughts. When we behold what is to be seen upon the Earth, not from a human body but from the sphere of the fire spirits—that is, from the Saturn nature that projects into the Earth, as it were—then we gain exactly the picture of the evolution of the Earth that I have described in *An Outline of Esoteric Science*.[3] This book is so composed that the thoughts appear as the thought content of the

---

3. *An Outline of Esoteric Science*, translated by Catherine Creeger, Anthroposophic Press, Great Barrington, MA, 1997.

world, seen from the perspective of the fire spirits. These things have a deep and real significance in several ways.

Take the gnomes and undines: they exist in the world, one can say, that borders on human consciousness; they are just beyond the threshold. Ordinary consciousness is protected from seeing these beings, for the fact is that these beings are not all benevolent. The benevolent beings are, for instance, those that I described yesterday as working in the most varied ways on plant growth. But not all these beings are well disposed. And the moment one breaks through into the world where they are active, one finds there not only the well-disposed beings but the malevolent ones as well. One must first form a conception about which of them are well disposed and which are malevolent. This is not easy, as one can see from the way I must describe the malevolent ones. The main difference between the ill-disposed beings and those that are benevolent is that the latter are always drawn more to the plant and mineral kingdoms, whereas the the former are drawn to the animal and human kingdoms. Some, which are even more malevolent, try to approach the kingdoms of the plants and the minerals as well. One can gain quite a fair idea of the malevolence of the beings of this realm when one turns to those that are drawn to human beings and animals, who, in particular, try to accomplish in human beings that which the higher hierarchies entrust to the well-disposed beings for the plant and mineral world.

There exist ill disposed beings in the realm of the gnomes and undines that approach human beings and animals and cause the complement with which they are supposed to endow only the lower animals to come to physical realization in human beings. This element is already present in humans, but the beings' aim is that it should be manifested physically in human beings as well as in animals. Through the action of these malevolent gnome and undine beings, animal and plant life of a low order—parasites— live in human beings and in animals. These malicious beings are the begetters of parasites. The moment humans cross the threshold of the spiritual world, they at once meet the trickery that exists in this world. Snares are everywhere, and we must first learn from the goblins to be attentive. This is something that spiritualists, for example, can never manage! Everywhere there are snares. Now someone might say: "Why, then, are there malevolent gnome and undine beings, if they engender parasites?" Well, if they did not exist, humans would never be able to develop within themselves the power to evolve brain matter. Here we encounter something of extraordinary significance.

If we consider the human being as consisting of metabolism and limbs; of the chest, or the rhythmical system; and then the head, or the system of nerves and senses, there are certain things about which we must be quite clear. Let us leave out the rhythmical sphere. Both below and above, certain processes are taking place. If we look at the processes taking place below, in the metabolic/limb system, as a whole, we find that in ordinary life they have one result that is usually disregarded. These processes are those of elimination, through the intestines, through the kidneys, and so on; all of them have their outlet in a downward direction. They are mostly regarded simply as processes of elimination, but this is nonsense. Elimination does not take place merely in order to eliminate; to the same degree to which the products of elimination form, an element arises spiritually in the lower sphere of the human that resembles the nature of the physical brain in the upper region. In the lower human being, a process is arrested halfway in its physical development. Elimination takes place because the process passes over into the spiritual. In the human being's upper sphere this process is taken to its conclusion. What below is only spiritual assumes physical form above. We have the physical brain above and a spiritual brain below. If one were to continue the transformation of this product of elimination, ultimately such metamorphosis would give rise to the human brain.

The human brain is the evolved product of elimination. This is of immense importance with respect to medicine, for instance, and it is something of which doctors in the sixteenth and seventeenth centuries were still fully aware. Of course, today people speak in derogatory terms—and rightly so—of the "quacks of old who dealt in filth," but this is because they do not know that such filth still contained "mummies" of the spirit. Naturally, this is not intended as a glorification of what has figured as quackery in past centuries, but I am referring to the many truths with connections as deep as those I have just cited. The brain is a higher metamorphosis of the products of elimination. Herein lies the connection between diseases of the brain and intestinal diseases, and also their cure. Because gnomes and undines exist, because there is a world in which they are able to live, there are forces that are capable of giving rise to parasites in the human being's lower sphere; at the same time, in the upper sphere, their presence allows the products of elimination to metamorphose into the brain. It would be absolutely impossible for us to have a brain if the world were not so ordered that gnomes and undines can exist.

What is true of gnomes and undines with regard to destructive powers—for destruction and disintegration also proceed from the brain—holds good for sylphs and fire spirits with regard to regenerative powers. The well-disposed sylphs and fire spirits keep away from human beings and animals and busy themselves with plant growth in the way I have indicated, but there are also those that are malevolent. These ill-disposed beings are most concerned in carrying what has its true place up above—in the regions of air and warmth—down into the watery and earthy regions. If one wishes to study what happens when these sylphs transmit what belongs up above down into the watery and earthy regions, look at the deadly nightshade (*Atropa belladonna*). The flower of this plant has been kissed by the sylph, as it were, and potentially beneficent juices have been changed into poisonous ones.

Here we have what may be called a displacement of spheres. It is appropriate for the sylphs to develop their enveloping forces up above, where the light plays over one, for the bird world needs this. If, however, the sylph descends and makes use in the plant world of what it should employ up above, a potent vegetable poison is engendered. Sylphs give rise to the poison representing a heavenly element that has streamed down too far, that has descended to Earth. When people or certain animals eat the fruit of the deadly nightshade, which looks like a cherry except that it lies concealed in the calyx, it is fatal to them. But just look at the thrushes and blackbirds; they perch on the plant and get from it the best food in the world. What is present in belladonna belongs rightly to the sphere they inhabit.

This is remarkable. Animals and human beings, whose lower organs are earthbound, experience as poison a substance that has become corrupted on the Earth in the deadly nightshade. But the birds, which should obtain this substance in a spiritual way from the sylphs—and who, through the benevolent sylphs, in fact do so obtain it—are able to assimilate this substance that has been thrust down to the Earth from their region above. They find nourishment in what is poison for those beings that are more bound to the Earth. Thus one can gain a conception of the way in which gnomes and undines impel something of a parasitic nature to strive upward from the Earth toward other beings, while poisons filter downward through the action of the sylphs.

When the fire spirits imbue themselves with the impulses that belong in the region of the butterflies—and are of great use to them in their development—and carry those impulses down into the fruits, we get the poisonous almond species. The poison is carried into the fruit of the almond trees through the activity of the fire spirits, and yet the fruit of the almond could not come into existence at all if those same fire spirits did not in a beneficial way burn, as it were, the edible part of other fruits. With other fruits one has the white kernel at the center and around it the flesh of the fruit. With the almond one has the kernel in the center, and around it the flesh of the fruit is quite burned up. That is the activity of the fire spirits. If this activity miscarries; if what the fire spirits work to bring about is not confined to the brown shell, where their activity can still be beneficial; and if part of what should be engaged in developing the shell penetrates into the white kernel, then the almond becomes poisonous.

Here we have gained a picture of the beings lying immediately beyond the threshold of normal awareness and of how, if they follow their impulses, they become the bearers of parasitic and poisonous principles and therewith of illnesses. It becomes clear how, when we are healthy, we raise ourselves above the forces that take hold of us in illness. For illness springs from the malevolence of the beings necessary for the regeneration of the world of nature, for its shooting and sprouting growth but also for its fading and decay. These are the things that, rooted in instinctive clairvoyance, underlie such intuitions as those of the Indian Brahma, Vishnu, and Shiva. Brahma represents the beings at work in the cosmic sphere, which may legitimately approach the human being. Vishnu represents the cosmic sphere, which may approach humans only insofar as what has been built up must again be broken down, must be continually transformed. Shiva represents all that is connected with the forces of destruction. In the ancient times of the flowering of Indian civilization, it was said: "Brahma is intimately related to all that is of the nature of the fire spirits and the sylphs, Vishnu with all that is of the nature of sylphs and undines, Shiva with all that is of the nature of undines and gnomes." Generally speaking, when we look back to these more ancient conceptions, we find everywhere the pictorial expressions for what must be sought today as immanent in the secrets of nature.

Yesterday we studied the connection of these invisible folk with the plant world; today we have added their connection with the world of the animals. Everywhere, beings on this side of the threshold interact with those from beyond it and vice versa. Only when one knows the living interaction of both these kinds of beings does one really understand how the visible world unfolds. Knowledge of the supersensible world is indeed very necessary for humans. The moment we pass through the gate of death, we no longer have the sense perceptible world around us, but from the beginning this other world begins to be our world. At our present stage of evolution, we human beings cannot find the proper access to the supersensible world unless we recognize in physical manifestations the written characters that point to this world. We cannot approach the supersensible if we have not learned to read in the creatures of the Earth, in the creatures of the water, in the creatures of the air, and in the creatures of the light the signs of elemental beings that are our companions between death and a new birth. We see of the creatures here on Earth between birth and death only their crude, dense parts. We learn to recognize their supersensible nature only when, with insight and understanding, we enter into this supersensible world.

5

Agriculture I

A SPIRITUAL SCIENTIFIC APPROACH to agriculture places the natural world in a larger context, considering both its material and its spiritual aspects. Modern science takes a more narrow perspective, focusing on ever-smaller and more limited fields of study. It does not consider each detail in a greater framework to find the wider interrelationships. With respect to the food we eat—the products of agriculture—its value does not lie strictly in the minute physical composition of the substances so much as in the vitality it gives to human beings, a vitality that can give us the capacity to develop spiritually. Plants will take from properly prepared soil the substances necessary to their own vitality, and human beings who eat the plants will likewise partake of these revitalizing forces. It is important, then, to reflect carefully on the environment in which plants grow and the ways in which we can aid the forces of nature (the nature spirits). We must think first about the soil in which a plant develops and how to enrich it properly. Two biodynamic spray preparations can be used to begin this re-enlivening process. If we place cow's manure in the horn of a cow and bury it in the Earth for the winter, when the Earth is most active in its depths, the astral and life forces will become concentrated. Spraying a diluted form of this preparation over the plowed fields in the spring encourages root growth and the formation of humus. Conversely, we also fill a cow horn with quartz and bury it for the duration of the summer, where it will be played upon by warmth and light. The following spring, we have a substance that complements the action of the cow horn manure. When sprayed on plants, this mixture fosters the growth of seed and fruit. These two preparations act in concert to push the plant upward out of the soil and draw it out into full flower.

LECTURE BY RUDOLF STEINER[1]
KOBERWITZ, JUNE 12, 1924

We have seen that in order to arrive at spiritual scientific methods applicable to agriculture, we need to look at nature and the spirit's activity in nature in its entirety, in its most encompassing dimensions. Materialistic scientists, on the other hand, have increasingly tended to narrow their scope and investigate ever more minute entities, although with agriculture, at least, they have not immediately gotten down to the microscopic level, which is so often the focus in other scientific disciplines. Even in agriculture, however, they still tend to be concerned with very small spheres of activity and the conclusions that can be drawn from them. But it is impossible to assess the world of human beings and other living things solely from such narrow perspectives. The way in which current science deals with the realities of agriculture is equivalent to trying to reconstruct the totality of a human being from just a little finger and an earlobe. Today there is an absolutely urgent need to counteract this tendency with a genuine science that can encompass large-scale cosmic interrelationships.

---

1. Lecture 4 from *(Agriculture) Spiritual Foundations for the Renewal of Agriculture*, eight lectures, Koberwitz, June 7–16, 1924 (GA 327).

Just think of how often current science—or the science of a few years ago—has had to correct itself. Take, for example, the scientific absurdities that were prevalent not too long ago regarding human nutrition. Everything was highly scientific and scientifically proven, and from a certain perspective the proofs were irrefutable. It was scientifically proved that an average person weighing 70–75 kilograms [155–165 pounds] required about 120 grams [4.2 ounces] of protein a day for adequate nourishment. As I said, this was "scientifically proved," yet today no scientifically inclined person believes this anymore. Science has corrected itself. Today everyone knows that eating 120 grams of protein per day not only is not necessary but is downright harmful and that people will be healthiest if they consume no more than 50 grams [1.8 ounces] a day. We know that excessive protein consumption creates by-products in the intestines that are, in fact, toxic. If one considers an entire lifetime and not just the period of life when the protein is being eaten, one will find that arteriosclerosis in old age is primarily due to the toxic effects of this excess protein. This is an example of how scientific investigations involving human subjects are often flawed because they consider only short-term effects. A normal human life span, however, is much longer than ten years, and although initial results may seem favorable, detrimental effects often become evident much later on.

Spiritual science is less likely to fall into errors like these. It is certainly not my intention to join in the cheap criticism that is often leveled at current science because it has had to adjust itself as I described; such things cannot always be helped. On the other hand, it is just as cheap to criticize spiritual science when it enters into practical life and is obliged for once to consider life's larger relationships and to investigate not just the forces and substances that are crudely material but also those that are more spiritual. This certainly applies generally to agriculture and to the issue of fertilizers in particular. Even the way scientists phrase their statements about fertilizing shows that they do not have any idea of its significance in the household of nature. We often hear that "fertilizers contain the nutrients for plants." I deliberately said what I did just now about human nutrition in order to show how even in very recent times, science has had to modify its views. It has had to do so because it started from a totally false view of how living things are nourished.

People have believed that the most important factor in nutrition—please do not take it amiss if I speak quite freely now—is what we put into our mouths and eat. Certainly, what we eat is important, but most of what we consume on a daily basis is not meant to be absorbed into the body and deposited as substance. It is meant to convey its forces to the body, to rouse the body to activity. Most of what we consume is excreted again, so that with regard to metabolism, the important thing is not so much the correct weights and proportions but whether or not the vitality of the forces in the food can be taken up in the right way. We need this vitality, for instance, whenever we walk or work or even simply move our arms.

On the other hand, the human body takes up through the sensory organs, through the skin, and through respiration most of the substance with which it enriches itself and which is then expelled as the body renews its substance every seven or eight years. What the body takes in and retains as substance is continually being absorbed in extremely fine doses; it becomes condensed only once it is inside the body. The body absorbs it from the air and then hardens and condenses it to the point where we eventually cut it off in the form of nails, hair, and so forth. It is completely incorrect to hold to the formula that we ingest food, and then it passes through the body and sloughs off in the form of hair, nails, skin, and so forth. In fact, the process is breathing and refined uptake through the sensory organs (even the eyes), passage through the organism, and then sloughing off. On the other hand, what we take in through the stomach is important, because it has inner energy like a fuel, because it brings in the forces that enable the will to be active in the body.

This is the truth, the simple result of spiritual research, and it can drive one to despair to see modern science defending exactly the opposite point of view. I say this because when it comes to the most important questions, it is so very difficult to come to any kind of understanding with modern science—and yet we must come to terms with it. If we do not, today's science will soon arrive at a dead end as far as practical matters are concerned, since its own methods prevent it from understanding some things that are right under its nose. I am not talking about experiments; as a rule, what science says about experiments is true, and they are very useful. It is the theorizing after the fact that is so deficient, for from this theorizing derive the practical guidelines in many different fields of activity. From this perspective, one can see how difficult it is to come to any kind of true understanding. On the other hand, we must reach such understanding, especially in the practical areas of life, which include agriculture. If we want to manage the various domains of agriculture properly, it is necessary to gain insight into the way substances and forces work and also into the way the spirit works. Just as a small child who does not yet understand the purpose of a comb may chew on it or misuse it in some other way, so, too, will we misuse things if we do not understand their essential nature.

To understand what I mean, let us begin by considering a tree. A tree is different from an ordinary annual plant that never grows past the herbaceous stage. A tree surrounds itself with bark and so on, but what is the essential nature of a tree in contrast to an annual plant? Let us compare the tree with a mound of soil that is uncommonly rich in humus, that contains a large amount of plant matter in the process of decomposition and perhaps some decomposing animal products too. (Steiner makes a drawing on the blackboard.)

On the left side is the mound of humus rich soil, in which I have made a crater shaped depression. On the right is the tree, which is more or less solid on the outside and which, on the inside, is growing and creating its shape. I am sure that it seems strange that I have drawn these two side by side, but they are actually much more closely related than one might think. Soil that is permeated with humus-like substances in the process of decomposition contains living ether. That is what is important. When such soil shows us, by virtue of its particular structure and character, that it contains matter that is etheric and alive, it means that the soil is on the way to becoming a kind of plant sheathing. It does not go so far as to become the kind of sheathing that makes up the bark of a tree. As one might imagine, it cannot evolve that far in nature. Still, at a higher stage of evolution this mound that incorporates the special etheric qualities characteristic of humus simply wraps itself around the plant. For any given locality on Earth, there is a certain level that separates what is above the Earth from what is inside the Earth. Anything raised above the normal level for that locale will show a particular tendency to life, a propensity to become permeated with etheric vitality. One will therefore find it easier to permeate ordinary inorganic soil with humus-like material—or any other kind of refuse in the process of decomposition—if one first builds up the soil into mounds. Then, of its own accord, the soil will tend to become inwardly alive and plant-like. The same process takes place in the formation of a tree. The soil mounds up and surrounds the plant; it envelops the tree with its etheric vitality.

Thus one will begin to comprehend that there is an intimate relationship between the content within the contours of a plant and the soil surrounding the plant. It is not at all true that life stops at the plant's perimeter. Life as such carries on from the roots of the plant into the soil, and for many plants there is no sharp dividing line between the life inside them and the life in their surroundings. We must be imbued with this idea and understand it thoroughly in order to appreciate the nature of soil that has been manured or otherwise fertilized in some way.

Fertilizing must consist of enlivening the soil, so that the plant is not growing in dead soil and does not have difficulty reaching the fruiting stage out of its own vitality. The plant does this more easily if from the very beginning it is embedded in living matter. All plant growth has a slightly parasitic quality, because it grows out of the living soil like a parasite. In many regions of the world we have to help plants by fertilizing them, because we cannot count on nature itself to incorporate enough organic residues into the soil and break them down far enough so that the soil becomes sufficiently enlivened. This is least necessary in the regions where there is so-called black earth;[2] there nature itself sees to it that the soil is sufficiently alive.

We must truly comprehend the nature of fertilization, but we must also understand how to establish a kind of personal relationship to the elements of farming, especially to the various manures and the methods of working with them. This may seem unpleasant, but without a personal relationship, we cannot succeed. Why not? The reason will soon become evident if one considers the fact that living things always have an inner and an outer aspect. The inner part is enclosed within a skin of some kind, and the outer part is outside that skin. Let us consider the inside.

The inside of a living organism has not only forces that stream outward, but also forces that stream inward, that are turned back. In addition, the organism is surrounded by all kinds of forces. Now there is a way to express very exactly, if rather personally, how a living organism establishes a balance between its inner and outer aspects. All the forces working to stimulate and maintain life within the living organism—within the contours of its skin—have to smell; one could even say they have to stink. This is what it means to be alive, that the substances that would emit odors if they were released are held together, that they do not stream outward too strongly but are held together inside and smell there. An organism lives in such a way that it allows as little as possible of

---

2. The regions of black earth—also called chernozem or mollisol—are found primarily in the Great Plains of North America and in the steppes of Russia.

the odor generated by the life within it to slip past the boundary of its skin. One might say that the healthier an organism is, the more it smells on the inside and the less it smells on the outside. A plant is not predisposed to give off odors, but rather to absorb them. If we comprehend the beneficial effects of an aromatic meadow full of fragrant plants, we realize the kind of mutual support that takes place among living things. The fragrance that wafts around and about works on the plant from outside. It is different from the mere life scent, and it spreads for reasons that we will examine later. We must have a personal relationship to these matters; only then will we truly participate in nature.

It is important to realize that manuring and similar practices must consist in providing the soil with a certain degree of vitality. In addition, they must allow the soil to develop the process I called special attention to yesterday. Nitrogen must be able to permeate the soil in such a way that life is guided toward certain lines of force, as I illustrated. Thus, when we fertilize, we must bring sufficient nitrogen to the earth realm so that in the fields where plants are to be grown, those structures in the earth realm that need life are imbued with it. That is the task, and it must be carried out in a sober and exact manner.

This description could lead us to suspect that when we use pure minerals as fertilizer, we can never really influence the earthy element in the soil but only, at most, the watery element. By using mineral fertilizers, one can influence the soil's watery element to some degree, but one will never be able to enliven the earthy element itself. Plants that have been treated with any kind of mineral fertilizer will show by their growth that they have been sustained only by stimulated water and not by enlivened earthiness.

If we want to understand these matters, we would do best to turn first to the least pretentious type of fertilizer, namely, compost, which is sometimes treated with disdain. In compost we have a means of enlivening the soil; compost can include any kind of organic waste from the farm or garden—spoiled hay, fallen leaves, even dead animals. We should not turn up our noses at things like this, since they still contain valuable etheric forces and also astral forces. These forces are not as strong as those in solid manure or liquid manure but in a certain sense they are more stable; the astrality, in particular, is more stable. It is simply a matter of supporting this stable astrality in the right way. If the etheric life in a compost pile is too rampant, the ability of the astrality to influence the nitrogen content is immediately impaired; the astrality is prevented from coming into its own.

There is a particular substance—lime—whose favorable influence in the natural world I have already described from several different points of view. If one adds lime to the compost pile, perhaps in the form of quicklime, a remarkable thing happens. Without causing the astrality to volatilize too strongly, the quicklime takes up the etheric, along with oxygen, so that the astrality becomes effective in a wonderful way. When one uses this compost as fertilizer, one is imbuing the soil's earthy component with something that strongly permeates it with astrality, without detouring through the etheric. Just picture it: astrality penetrates very strongly into the earthy component of the soil—without first going by way of the etheric—so that the soil becomes strongly astralized and by means of what is astralized becomes permeated with nitrogenous substance.

What arises is very similar to a certain plantlike process in the human organism, a plantlike process that is not concerned with reaching the fruiting stage but is content to remain, as it were, at the stage of leaf and stem formation. The process we impart to the soil must also exist in us, so that in a corresponding way we can bring our food to the necessary energetic state I mentioned earlier. When we treat the soil in the way I have described, we stimulate it to the same vitality. This prepares the soil to bring forth something that is especially good for animals to consume, so that under its influence they can develop inner liveliness, their bodies can become inwardly energetic. In other words, we would do well to fertilize our pastures and meadows with this compost. If we do so conscientiously—that is, if we also carry out other necessary procedures—we will end up with good pasturage, which even when it is cut will make good hay. In order to proceed in the appropriate way, however, we need to have insight into the whole matter. What we do in particular instances will naturally depend on our feeling, but this feeling develops once we have proper insight into the overall nature of the process.

If one simply leaves a compost pile lying around as I have described it so far, it can very easily begin to scatter its astrality. At this point the personal relationship one develops to this process becomes important, because one needs to do whatever one can to make the pile smell as little as possible. This can be easily accomplished if one builds up the pile in thin layers and covers each layer with loose peat or a similar substance before adding the next layer. This holds together the nitrogen, which wants to escape in all kinds of guises and combinations and would otherwise escape in the form of odors. The main point is that we must approach all aspects of farming with the conviction that for the whole process to work, we need to pour life and also astrality into everything around us.

Consider the question of why cows have horns, but certain other animals have antlers. That is an extremely important question. However, what science has to tell us on this point is usually very one-sided and superficial. Let us try to answer this question of why cows have horns. Remember that I said that a living organism need not have forces that stream only outward; it can also have forces that stream inward. Imagine an organic entity with streams of force traveling in both directions. If that were all there was to it, we would have a rather irregular, lumpy creature. A cow would look very peculiar. Its limbs would remain tiny, the way they are at the earliest embryonic stages. It would simply look grotesque. But that is not how a cow is put together; a cow has horns and also hoofs. What happens at the regions of the body where the horns and hoofs grow? There the streams are especially strongly turned inward, and the outer world is particularly strongly shut off. All communication with the outer world, such as can occur through skin or hair, is completely ruled out. In this way, the development of the horns and hoofs is connected to the form and development of the animal as a whole.

Antler formation is a totally different process. In the case of antlers, the streams are not directed back into the organism; instead, the antlers serve as outlets so that certain streams can be led outward and discharged. ("Streams" do not always have to be fluid or gaseous; they can also be streams of force, which in this case are localized in the antlers.) A stag is beautiful, because it stands in intense communication with its surroundings and directs some of its force streams outward; it lives at one with its environment and thereby takes in everything that influences the nerves and senses at an organic level. A stag is a quick and nervous animal. In a certain respect, all animals with antlers are suffused with a slight nervousness, which can be seen in their eyes.

A cow has horns in order to send the formative astral etheric forces back into its digestive system, so that much work can be accomplished there by means of these radiations from the horns and hoofs. Anyone trying to understand hoof-and-mouth disease—that is, how the periphery of the animal works back on the digestive tract—needs to appreciate this relationship. Our remedy for hoof-and-mouth disease is based on insight into this relationship. There is something inherent in a horn that makes it well suited for reflecting living and astral influences back into the activity of the interior. A horn can radiate life and even astrality. It is truly so. If one could crawl around inside the living body of a cow, right inside the cow's belly, one would be able to smell how living astrality streams inward from the horns. It is similar with the hoofs.

This points to measures we might introduce to enhance the effectiveness of ordinary barn manure. What is this manure in reality? It is outer nourishment that has entered and been absorbed into the animal to some extent and has permitted the dynamic development of forces within the organism but which then has been excreted, rather than serving primarily to enrich the animal with substance. In passing through the organism, however, it has become pervaded with astral and etheric activity. It has become permeated with nitrogen-bearing forces and oxygen-bearing forces. The mass that emerges as manure is impregnated with these forces .

Suppose we take this mass in some form and give it over to the Earth. We would be putting etheric and astral forces into the Earth, which by rights belong in the belly of an animal and which there produce plantlike forces. Indeed, the forces that are engendered in our own digestive tracts are plantlike. We ought actually to be very grateful that there is anything left over in the form of manure, because through it etheric and astral forces are carried from inside the organs out into the open. These etheric and astral forces are inherent in the manure; all we have to do is to preserve them so that they can have an enlivening and astralizing effect on the soil—not just on the watery part of the soil but also on the earthy, mineral component. Manure has sufficient force to overcome even the inorganic earth element.

The substance we put into the ground, of course, must lose the form it originally had before it was consumed as fodder; it must first pass through an internal organic process in the animal's metabolic system. To a certain extent the resulting substance will have begun to decay and dissolve, but it is best if it has just reached the point where it is starting to fall apart by virtue of its own etheric and astral forces. Then microscopic parasites, the smallest living creatures, put in an appearance and find a fertile breeding ground in the manure. People think that these parasitic beings have something to do with the manure's quality. They are only a sign, however, that the manure is in a certain condition. Bacteria can be useful as indicators, but we are succumbing to an illusion if we think that we can radically improve manure by inoculating it with bacteria and so on. This might initially appear to be true, but in reality that is not the case. I will return to this later.

Next let us take the manure in whatever form is available, stuff it into a cow horn, and bury it in the ground at a certain depth—

I would say between one half and three quarters of a meter [18 inches to 38 inches], provided the soil beneath is neither too clayey nor too sandy. By burying the cow horn filled with manure, we preserve in the horn the etheric and astral forces that the horn was accustomed to reflect back when it was still part of the cow. Because the cow horn is now outwardly surrounded by the Earth, all the Earth's etherizing and astralizing rays stream into its inner cavity. The manure inside the horn attracts these forces and is inwardly enlivened by them. If the horn is buried for the entire winter—the season when the Earth is most inwardly alive—all this life will be preserved in the manure, turning the contents of the horn into an extremely concentrated enlivening and fertilizing force. Once winter is over, we can dig up the cow horn and take the manure out of it. During our last experiments in Dornach, people discovered that when we took the manure out, it did not stink at all anymore. This was very striking. It had no smell, although of course it did begin to smell a little as we mixed it with water. This shows that all the smell had been concentrated and transformed. After spending the winter underground, the cow horn contains an immense astral and etheric energy, which one can use by diluting the contents with ordinary water, perhaps warmed up a bit.

I have always found—inasmuch as I first look at the area to be treated to get an impression of the quantities—that for an area as large as, say, the distance from the third window here to the first footpath, we need only one hornful of manure diluted in about half a bucket of water. We must make sure, however, that the entire contents of the horn have been thoroughly exposed to water. We have to start stirring it quickly around the edge of the bucket, on the periphery, until a crater forms that reaches nearly to the bottom, so that it is rotating rapidly. Then reverse direction quickly, so that the mixture seethes and starts to swirl in the opposite direction. If we continue doing this for an hour, the mixture will be thoroughly blended. Just imagine how little work this takes! The burden of labor is not very great. Moreover, I can imagine that otherwise unoccupied members of a farming household might really enjoy stirring the manure, at least for a while. The sons and daughters of the household could do an excellent job of it. It is a very pleasant feeling to experience how a delicate odor develops out of what was initially odorless. The personal relationship that we can develop to such processes is extremely beneficial for people who wish to experience nature as a whole and not just as it appears in the Baedecker guidebooks.

The next thing to do is to spray the substance over plowed fields, so that it can really unite with the soil. When only small areas are involved, one can apply the mixture with an ordinary syringe sprayer; for larger areas, special spraying machines will have to be used. If a person manages to supplement the usual manuring with this kind of "spiritual manure," one will soon see the fertility that will result. In particular, one will find that this type of fertilizing lends itself to further development in a most remarkable way. We can follow up the measure that I have just explained with another one, also using the cow's horn.

This time, instead of stuffing the cow horn with manure, fill it with quartz that has been ground to a powder and mixed with water to the consistency of a very thin dough. Instead of quartz, one could also use orthoclase, a type of feldspar.[3] Instead of leaving the cow horn in the ground through the winter, let it remain underground all summer and take it out in late autumn. Save the contents, which have been exposed to the summer life of the Earth, until the next spring and treat them just as one did the manure. In this case, however, one will need much smaller quantities; one can take a portion the size of a pea or perhaps no bigger than a pinhead and stir it into a whole bucket of water. This mixture, too, needs to be stirred for an hour. If one applies this as a fine spray on the plants themselves—especially on vegetables

---

3. Feldspar refers to a group of aluminum silicates, principally of soda, potash, and lime; orthoclase is potassium aluminum silicate ($KalSi_3O_8$).

and the like—one will soon notice how the mixture's effect complements and supports the influence coming from the other side, from the soil, as a result of the cow's horn manure. If one were to extend this treatment to entire fields, one would then see how the cow horn manure pushes up from below and how the other substance pulls from above, neither too strongly nor too gently. The combination would be wonderfully effective, especially in the case of grains.[4]

These methods are derived from a wider field of observation and not just from dealing with a single question, which is like trying to reconstruct a whole human being from one finger. We should not underestimate what is thus attained. The studies that are made today—on "farm productivity" and the like—are really nothing more than studies on how to make production as profitable as possible. They do not amount to much more than that. Of course, even farmers are impressed, although they are not always conscious of it, when some measure or other achieves big results—huge potatoes, for example, or any other crop that is big and swollen. But then the investigation is not taken any further, even though the visible results are not the most important aspect at all. More important is that once these crops reach human beings, their effect on human life should be as beneficial and health giving as possible. We can raise fruits that look magnificent in the orchard or field, but as far as human beings are concerned, they may simply be stomach filler and may not promote human life. But today's science is unable to pursue matters to the point where people would be deriving the best kind of nourishment; it simply does not know how to go about it.

In contrast, what spiritual science has to say on the subject is rooted in the whole household of nature. It is always conceived out of the totality, and for this reason the individual measures are

---

4. The horn manure and horn silica sprays mentioned in this lecture are also known respectively as Preparations 500 and 501.

also effective for the totality. If we farm in this way, the result can be nothing other than what is best for human beings and also for animals. In spiritual science, human beings are the standard and starting point. All our practical suggestions serve the purpose of sustaining the whole human being in the best way possible. That is what distinguishes this kind of study and research from what is customary today.

DISCUSSION

(QUESTION) *Does the dilution [of the horn manure preparation] continue in arithmetic progression?*

(DR. STEINER) That is something that needs to be investigated further. You will probably find that as the areas to be treated get larger, you can use greater amounts of water and fewer cow horns. Thus, you will be able to manure large areas with relatively few cow horns. In Dornach, we had twenty five cow horns, and we used some of them to treat a large garden. First we used one horn to half a bucketful of water; later we treated it again, using two horns to a whole bucket. Then we also had to treat an area that was much bigger; there we used seven horns to seven buckets of water.

Is it acceptable to use a machine to stir the manure for larger areas, or is that not permissible?

You can either be quite strict about these matters, or you can decide to gradually slide toward surrogates. There is no question that stirring by hand has a quite different significance than does mechanical stirring, though someone with a mechanistic worldview would never admit it. Consider what a huge difference there really is: when you stir by hand, all the fine movements of your hand go into the stirring; quite possibly all kinds of other things do, too, including the feelings you have as you stir. People today do not think that makes any difference, but in the field of medicine, for instance, the difference is quite noticeable. Believe me, it is really not a matter of indifference whether a certain medication is prepared by hand or by machine. Something is imparted to that which is produced by hand—you must not laugh at this.

I have often been asked what I thought of the Ritter remedies,[5] about which some of you may have heard. As you may know, some people sing the praises of these remedies, while others maintain that they have no particular effect. They do indeed have an effect, but I am firmly convinced that they would lose a great deal of their effectiveness if they became available commercially. With these medicines in particular, it makes a great difference when the doctor prepares the remedy and gives it directly to the patient. A certain enthusiasm is administered along with the remedy when administration takes place only in smaller circles. You may say that enthusiasm cannot be weighed or measured, but an enthusiastic doctor is an inspired doctor, and the doctor's enthusiasm supports the dynamic effect of the medicine. After all, light has a strong influence on remedies, so why shouldn't enthusiasm also? As a mediator, enthusiasm can have a very great effect, and so can an enthusiastic doctor. That is why the Ritter remedies can be so effective. Great things can be accomplished with enthusiasm. However, if a substance were to be prepared by machine instead of the human hand, the effects would probably disappear. You might also discover that the stirring is such fun that you would not even consider using a machine, even when many cow horns are needed. It could be something you do on Sundays after lunch. If you invite a lot of guests and provide some entertainment, you could get it done just beautifully without any machines.

---

5. Marie Ritter (d. 1924) was a German pharmacist who developed "photodynamic" herbal remedies (not related to biodynamic agriculture).

There already are technical difficulties in distributing half a bucket of water over one-fifth of an acre. If the number of cow horns is increased, this would lead to more difficulties in the distribution. Can that half bucket of water be diluted with more water, or is it important to keep this proportion? Should we always use about half a bucket for one-fifth of an acre?

You could do as you suggest, but then I think the method of stirring would have to be modified. You could finish stirring one hornful in a half a bucket of water and then further dilute what is in the bucket, but then you would have to stir it again. I think it would be better to calculate how much less of the material you would need in a half bucket of water and then stir by the half bucketful, even if you use less than one hornful. It is very important to get a thorough and intimate mixture; simply pouring the stuff into the water and stirring it will not do at all. You must bring about an intimate permeation, and this does not happen when you put in any kind of thickened substance or if you do not stir vigorously. I think it is easier for people to stir a lot of half buckets with a little bit of substance in them than to stir the diluted water a second time.

Since there will always be some solid particles left, could the liquid be strained so that it can be better applied with a spraying apparatus?

I do not think that is necessary. If you stir it quickly, you will get a fairly cloudy liquid and will not have to worry about there still being foreign particles in it. The manure will allow itself to be evenly distributed. Pure cow manure is best, but I do not think that you have to go to the extra effort of straining it, even if there are other substances mixed in. Rather than being harmful, the other material might even be beneficial, because the process of concentration and subsequent dilution involves only a dynamic radiant effect, not a substantial one. Thus, if a foreign body lands somewhere, there is no danger, for instance, of getting potato plants with long shoots and nothing on them. There is no danger of that.

I was only thinking of what it would do to the sprayer.

You can put it through a strainer; that will not hurt. The best thing would be to have your machinery fitted with a strainer in front of the sprayer.

You did not say whether we should weigh what comes out of the horn so as to get a specific proportion. And is that half bucket a Swiss bucket, or is it a certain number of liters?

I took an ordinary Swiss milk bucket—everything was developed from direct experience. The weights and measures need to be worked out.[6]

Can the cow horns be reused, or do they always have to come from freshly slaughtered animals?

We have not tested it, but as far as I can tell, you ought to be able to use the horns three or four times. After that, they probably will not work as well. It might be possible to use them for an extra year if you store them in the cow barn for a while after they have been used three or four times. I do not have any idea, though, how many horns will be available on the average farm, so I cannot tell whether you will need to be especially economical with them. That is a question I cannot answer right now.

Where can we get the cow horns? Do they have to have come from Eastern or Central Europe?

Where you obtain them does not matter at all. They must not come, however, from slaughterhouse refuse—they need to be as fresh as possible. However, the strange thing is that life in the Western Hemisphere is something quite different from life in the Eastern Hemisphere. Life in Africa, Asia, and Europe has a totally

---

6. A Swiss milk bucket holds about eight liters (2 gallons). Recommended weights and measures vary for specific countries or climates; information can be obtained from local Bio-Dynamic associations.

different significance from life in America. It is possible that horns from American cattle would have to be treated somewhat differently to make them effective. For instance, you might have to pound the manure in more tightly to make it denser. It is best to use horns from your own locality. There is a very strong kinship between the forces present in the cow horns of a given area and the other forces at work in the region; the forces of the foreign horns might conflict with the local forces of the Earth. You also have to take into account that frequently the cows providing the horns in a given place do not originate in that area. This is not much of a problem, because when cows have been grazing and living on a certain piece of land for three or four years, they then belong to that land, unless they are Western cattle.

How old can the horns be? Do they have to come from an old or a young cow?
   You will have to experiment; from what I know, it would seem that horns from cows of middle age would be best.

How big do the horns have to be?
   [Rudolf Steiner drew the size of the horn on the blackboard—about 12 to 16 inches—the average horn size of the local Allgäu cattle.]

Is it also important whether the horn comes from an ox or from a male or female animal?
   It is very likely that horns from castrated cattle would not work at all and that bull horns would not work very well. That is why I keep saying "cow" horns. Cows, as a general rule, are female! I mean the female animal.

When is the best time to sow grains for breadmaking?
   A more exact answer to this question will emerge when I talk about sowing later on in the lectures. Planting dates are extremely important, and it makes a great deal of difference whether you

sow closer to the winter months or earlier in the year. If you sow nearer wintertime, you induce a strong reproductive capacity; if you sow earlier, you engender a strong nutritive force in your grains.

Could we also distribute the cow horn manure by mixing it with sand? Does it make any difference if it is raining?

With regard to the sand, you can go ahead and try it. We have not tried it, but there is nothing to be said against it. The effect of rain would have to be tested. Presumably the rain would not make any difference; it could even establish everything more firmly. On the other hand, since there is such a strong concentration of forces involved, it is also possible that even the tiny impact of falling raindrops might disperse things too much. It is really a very subtle effect we are working with, so all these things have to be taken into account.

How can we protect the cow horns and their contents from harmful influences during storage?

The general rule is that removing the so-called harmful influences does more damage than simply leaving them alone. People are so eager to disinfect everything these days that they invariably overdo it. In the case of our remedies, for instance, we found that if we wanted to prevent any possibility of mold, we would have had to use methods that would also have inhibited the therapeutic effect. I do not have all that much respect for these "harmful influences"; they do not do much damage. It is better to leave things alone than to go to great lengths to keep them clean. We just covered the cow horns with pig bladders to keep the dirt from falling in. As for the horns themselves, I would not recommend cleaning them in any special way. You have to get used to the idea that dirt is not always "dirt." If you cover your face with a thin layer of gold, that is "dirt," and yet gold is not dirt. Dirt is not always dirt; in fact, sometimes dirt is precisely what acts as a preserver.

Should we take any special measures to help drive the seed as far as possible into chaos?

You could, but it is not necessary. If seeds are formed at all, a maximum of chaos is already present and needs no further support. Manuring is where support is needed. As far as seed formation is concerned, I do not think there is any need to enhance the chaos. If the seeds are fertilized at all, the chaos is already complete.

Will it not also be necessary to do something for the cosmic forces that are supposed to be retained until a new plant is formed?

This could be done, of course, by making the soil more silicious, since it is through silica that the actual cosmic factor is absorbed by the Earth and becomes effective. You could do that, but I do not think it is necessary.

How big should our experimental plots be?

You could experiment in the following way. Giving general guidelines is always relatively easy, but you will have to see for yourself what size works best in practice. In this case it will be relatively easy to set up experiments. Let us say you plant two experimental beds of wheat and sainfoin side by side. You will find that the wheat, which has a strong natural tendency toward seed formation, will be hindered in this respect if you add silica to the soil; with the sainfoin, on the other hand, you will find that seed formation is probably suppressed or perhaps simply delayed. When you want to research these matters, you can always compare the features of a grain like wheat with analogous features in sainfoin or some other legume. In this way you can set up very interesting experiments on seed formation.

Does it matter when the diluted substance is applied to the fields?

Undoubtedly it does matter. As a general rule, you can leave the cow horns in the ground until you need them. They will be none the worse for it, even if they were meant to be left over winter and stay in for part of the summer. If it is necessary to keep them somewhere else, you should make a crate and line it with a cushion of loose peat on all sides. If you keep the cow horns in the crate, the powerful concentration inside them will be maintained. On the other hand, it is not at all advisable to keep what you have already diluted with water. You have to stir it not too long before you use it.

If we want to use this treatment on winter grains, should we use the cow horns a whole quarter of a year after they have been taken out of the ground?

It is always best to leave them in the ground until you want to use them, although it does not really matter. If you use them in early autumn, you can let them stay in the ground until you need them. It will not harm the manure.

Won't the etheric and astral forces be lost if you use a machine that breaks the liquid up into a very fine spray?

Not at all. They are very firmly bound. In general, you do not have to be nearly as afraid that spiritual things will run away from you as you do with material things—that is, unless you chase them away to begin with.

How do you handle the cow horns with the minerals in them that have spent the summer under ground?

These are not at all harmed if they are taken out and stored somewhere; you can just toss them in a pile. The material that has been under ground during the summer will not be harmed. The sun can shine on the horns; it might even do them good.

Do you have to bury the horns in the particular location you want to manure afterward, or can you bury them all next to each other somewhere else?

It makes so little difference that you do not have to worry about it. In practice, it is best to look for a spot that has relatively good soil with some humus in it, a soil that is not too highly mineral. Then you can bury all the horns you need in one spot.

What about using machines on the farm? Some people say we should not use any machinery at all.

Well, that is a question that cannot be answered agriculturally. Given the state of affairs in society today, it is not really relevant to ask whether machines should be used; it is hardly possible to be a farmer today without using machines. Not everything you do is related to the most intimate processes of nature in the same way as stirring of manure. In cases where we should not approach one of these intimate natural processes with a purely mechanical process, nature itself sees to it that machines are not of much use. You cannot do much with a machine when it comes to seed formation—nature takes care of that. I do not think the question is actually very relevant. Today, how can you manage without machines? On the other hand, I would also like to say that we do not need to encourage machine mania on the farm. I am sure that people who always have to have the latest and best in machinery will do less well as farmers—even if the new machinery is an improvement—than they would if they went on using their old machines until they wore out. Strictly speaking, however, these are not really agricultural issues anymore.

Can you use the given quantity of cow horn manure mixed with water on an area half as large?

If you do, you will end up with rampant growth. If you use it in growing potatoes, for instance, you will get uncontrolled growth and sprawling stems, and what you want will not really develop at all. You will get what are known as rank patches. If you have rank patches, you have used too much manure.

What about fodder plants, where rampant growth is desirable, or a crop like spinach?

Even in those cases, I think you ought to use no more than a half bucket with one cow horn, as we did in Dornach on an area that is mainly vegetable garden. You will need much less for crops grown in larger fields. That amount is the optimum.

Does it matter whether you use cow manure or horse manure or sheep manure?

Cow manure is undoubtedly the best material for this procedure, but it might not be a bad idea to investigate whether horse manure could be used too. If you want to use horse manure for this purpose, it would probably be necessary to wrap the horn with some hair from the horse's mane, so that those forces can be brought into play, since a horse has no horns.

Should we manure before or after sowing?

The right time to spread manure is before sowing. This year we got around to it a bit late, and some of the manuring was done after sowing. We will see whether it does any harm. Naturally, though, you should spread it before sowing so that the soil will already be influenced by the manure.

Can we use the same horns for both the manure and the ground minerals?

Yes, you can do that, but you still will not be able to use them more than three or four times altogether. They lose their forces after having been used three or four times.

Is it important who does the work? Can it be anyone at all, or should it be an anthroposophist?

That is a good question. Of course, people will laugh if you raise a question like that today, but let me remind you that some people who grow flowers in their window boxes are very successful, while with others the flowers simply wither away. This is true. These things, which cannot be explained outwardly but which are inwardly quite clear, happen through the personal influence of the human being. They can come about simply because the person in question practices meditation, becomes prepared through meditative life. Indeed, when people meditate, they relate quite differently to nitrogen, which contains the imaginations. They put themselves in a position that enables all these things to be fully effective; and they are then in this position with regard to all plant growth.

These days such influences are not as clear as they used to be. There were indeed times when people knew that through undertaking certain practices they made themselves fit to care for growing plants. When this is not taken into consideration, other people's attitudes rub off. The delicate and subtle influences are lost if one is constantly among people who do not heed such things. That is why these influences are so easy to refute. I still hesitate to talk openly on this topic in any large gathering, because the circumstances of life today are such that it is very easy to refute these matters.

The question raised by our friend Stegemann at the farmers' meeting the other day in the Bockschen Saal in Breslau, about whether it is possible to combat parasites by means of concentration and similar exercises is actually a very ticklish one. There is no doubt that it can be done, if you do it right. If you were to establish a sort of festival, especially during the season when the strongest forces unfold and are concentrated within the Earth— the time between mid January and mid February—and undertake certain concentration exercises at this time, there could well be some effect.

As I said, it is a very ticklish question, but it can be answered in the affirmative. You just have to do it in harmony with nature as a whole. You have to know that engaging in a concentration exercise is totally different in midwinter than it is in midsummer. Certain folk sayings contain much that can provide valuable hints today. One of the many things I should have done in this incarnation was to pursue the idea I had as a youth of writing a kind of peasant philosophy that would portray the conceptual life of the peasants with respect to the world around them. It could have been very beautiful, and it would certainly have refuted the statement that farmers and peasants are stupid. A subtle wisdom would have emerged, a philosophy that would have expounded wonderfully on the intimate details of nature, down to the very way in which words were put together. It is really astonishing how much the peasants knew about what goes on in nature.

Today, it would no longer be possible to write this peasant philosophy; almost everything has been lost. Things simply are not the same as they were forty or fifty years ago. Back then you could learn so much more among the peasants than you could at the university. That was an altogether different age. You lived with the peasant farmers in the country, and it was a rarity for people with broad brimmed hats to come along and try to introduce the modern socialist movement. Now the world is quite different. The younger ladies and gentlemen among you have no idea how much the world has changed in the past thirty or forty years. Much of the beauty in the country dialects has been lost and even more of the cultural philosophy of the country people. Even the peasant almanacs had content that they no longer have. I remember almanacs printed on cheap paper but with the planetary signs painted in different colors. On the title page, the first thing you found was a bit of sugar candy you could lick whenever you used the book. That was one way of making the subject palatable!

When we need to manure larger areas, should we decide how many cow horns to use simply according to what feels right?

I would not recommend this method. I think we really have to be rational. I would advise you first to obtain the best results according to what feels right to you; then translate your results into figures, so that you have proper tables of calculation that people can use. I would say that those of you who think you can manure simply by intuition should go ahead and do so; in communication with others, however, we should not act as if the tables are not valuable. People today need to have these methods translated into calculable figures and amounts. We need cow horns for this work, but that does not mean we should become bull headed in advocating it! This is just the kind of thing that so easily arouses opposition. I would advise you to compromise and to respect the opinions of the world as much as possible.

Should quicklime be used in compost piles in the proportions that are usually recommended?

I am sure that the old method will prove to be a good one. You may have to modify it a bit, depending on whether you have moorland soil or sandy soil. With sandy soil you will need somewhat less quicklime, while with moorland soil you will require somewhat more, because of the acids that are formed.

What about turning the compost heap?

That is not bad for it. Once you have turned it, however, you should try to protect it by covering it with a layer of earth. Peat soil or peat mold is especially good as a layering.

What kind of potash did you mean when you said that, if necessary, it might be used in a transition stage?

Potash magnesia

What is the best way to use the manure that is left over after the cow horns are filled? Should it be spread on the fields in the fall so that it can experience winter there, or should it be stored until spring?

You must be clear that fertilizing with cow horn manure is not meant to replace manuring in the usual way—you have to go on doing that too. You must think of the new method of manuring as an enhancement of the effect of the manuring that you have been doing up to now.

6

Agriculture II

THE FORCES THAT SURROUND the plant as it grows also live in the soil. Over time, these forces can get used up; the soil can become drained of its revitalizing power. The living Earth slowly dies and cannot support proper plant growth. It is our task to bring health and life back to the Earth. The agricultural methods in general use lay store by animal manures populated with microorganisms that are thought to be beneficial to the manure. Then, too, ordinary mineral-type, inorganic fertilizers are much favored. These un-enlivened forms of fertilizing cannot fulfill the soil's true needs. Only when certain compounds are present in the correct proportions can the soil be regarded as entirely healthy and vital and capable of stimulating plants in their growth. Various substances can be added to manure to rejuvenate the soil and bring about the appropriate balance of compounds within it. Yarrow, chamomile, stinging nettle, and dandelion, when properly prepared within certain sheaths of animal organs and body parts over the course of months and at specific times of the year, acquire this capacity to benefit the soil in a variety of ways. Yarrow, with its sulfur content, draws potash, a potassium compound, into the soil. Chamomile brings calcium, a constituent of lime, to bear on plant growth. Stinging nettle, with its affinity for iron, stabilizes the soil's nitrogen content. Dandelion attracts silica to the soil.

Crumbled oak bark can contribute to the soil an organic form of calcium to ward off plant diseases. Valerian blossoms, pressed and diluted in water, stimulate the activity of phosphorus. Together, these biodynamic preparations work to reinvigorate the soil, replenish nutrients, and aid in organic decomposition.

LECTURE BY RUDOLF STEINER[1]
KOBERWITZ, JUNE 13, 1924

The purpose of the cow horn preparations I introduced yesterday is simply to improve and enhance manuring; we still need to go on manuring. Today, we will consider how this manuring can be done in accordance with the perspective we developed yesterday, namely, that the etheric vitality must be retained within the realm of the living, that it should never leave the realm of growth. That is why it is so important for us to recognize that the soil surrounding the growing plants' roots is a living entity with a vegetative life of its own, a kind of extension of plant growth into the Earth. Yesterday I even went so far as to show how we can imagine the transition between a raised up mound of soil, which possesses some internal vitality owing to its humus content, and the bark that surrounds a tree and seals it off from the outside. In modern times, people have lost all insight into the great interrelationships of nature. Thus, it is not surprising that they have also lost all insight into how the life that is common to plants and to the soil extends into the excretory products of life as we find them in manure. It was inevitable that insight into the forces of life would gradually disappear.

---

1. Lecture 5 from *(Agriculture) Spiritual Foundations for the Renewal of Agriculture.*

Spiritual science should not act out of a certain fanaticism as something turbulent and revolutionary with respect to the achievements of the modern world in the various fields of practical life. These achievements should be fully acknowledged. The only things that merit our active opposition are those that are so bound up with the modern materialistic worldview that they rest on totally false premises. We must acknowledge what is legitimate and simply complement it with what can flow from a living conception of the world. That is why I am not going to spend much time describing how manure, liquid manure, or compost should be prepared for use as fertilizer; a great deal has already been accomplished in this regard. Perhaps we will have time to say more on the subject during the discussion session this afternoon.

Let me start by saying that the idea that we are exploiting the land when we farm is quite accurate. In fact, such exploitation is unavoidable, because with everything we send off the farm and into the world, we are taking away forces from the soil and even from the air, which need to be replenished. Eventually, the manure must be treated in order for it to acquire the capacity to properly vitalize the depleted soil. It is here that various misconceptions have recently arisen as a result of materialistic ways of thinking.

For example, scientists of today study in great detail the activity of bacteria, the smallest living things. They credit these microorganisms with being able to condition the manure in the right way, and this has led to extremely ingenious and logical attempts to inoculate the soil. In most cases, this inoculation has had no lasting positive effect, since the reasoning behind it is roughly equivalent to thinking that a room is dirty because there are a lot of flies in it. The room is not dirty because of the flies; the flies are there because the room is dirty. And we will never get the room any cleaner by dreaming up all kinds of ways of either getting rid of the flies or increasing their numbers in the hope that they will eat up all the dirt. Not much can be achieved through methods like that. It is much more effective to tackle the dirt directly. Thus, when animal manures are used as fertilizer, the appearance of microorganisms must simply be regarded as resulting from the processes that are taking place within the manure. These creatures are useful as an indication of the condition of the manure, but there is little to be gained by inoculating the manure with them or breeding them on purpose; if anything, it might be more useful to combat them. When we are dealing with the living realm, which is so important for agriculture, it is always a matter of keeping to larger perspectives. As much as possible, we must avoid regarding these tiny creatures in an atomistic sense.

Of course, making a statement such as this does no good at all unless we can also show how things ought to be done. This has been stressed by other people, but knowing what is right is not the only matter of importance. A method may be perfectly correct, but if it is negative, it is often not possible to make progress unless positive principles can be put forward as well. As a general rule, when it is not possible to make positive suggestions, it is better to refrain from stressing the negative, since that only makes people angry.

A second result of materialistic ways of thinking is the idea of treating the manure with all kinds of inorganic compounds and chemical elements. Once again, experience has shown that this has no lasting positive effect. We must understand that if we try to improve manure with inorganic chemicals, we will succeed only in penetrating and enlivening the watery part of the soil. To grow sound and substantial plants, however, this is not enough, because no further vitalization proceeds from the water that seeps through the soil. We have to enliven the soil directly; this cannot be done with mineral fertilizers but only by means of organic material that has been conditioned to organize and enliven the solid earth itself. It is the task of spiritual science with respect to agriculture to indicate how this stimulus can be imparted to manure or liquid manure or any other sort of organic matter. Spiritual science always looks to the large-scale activities of life and ignores the microscopic realm and the conclusions drawn from this realm, which are of less importance. The task of spiritual science is to observe the macrocosm, the broadest dimensions of the workings of nature, and to understand these workings.

In the agricultural literature, we find various statements—derived from what people believe to be their experience—to the effect that nitrogen, phosphoric acid, lime, potash, chlorine, and iron are all important components of the soil if plants are to thrive. On the other hand, silicic acid, lead, arsenic, mercury, and even sodium compounds are said to be of value to plants only as stimulants. Such statements show that people are groping in the dark. It is fortunate—thanks to old traditions, no doubt—that people do not do the crazy things to plants that they would if they seriously followed such guidelines. As a matter of fact, it is impossible to follow them.

What, then, is the true state of affairs? If we fail to take proper account of silicic acid, lead, mercury, or arsenic, nature does not leave us as forsaken as it does if we fail to properly consider potash or lime or phosphoric acid. This is because silicic acid, lead, mercury, and arsenic are provided by the heavens, given freely along with the rain. On the other hand, for phosphoric acid, potash, and lime to be adequately imparted to the soil, we must properly manure the soil, since these substances are not a gift from heaven. Even so, over many years of farming, it is possible to deplete the soil. Indeed, we deplete it constantly—that is why we have to apply manure. But eventually, as is already the case on many farms, replacement through manuring is no longer sufficient. Then we begin to exploit the land, and it becomes permanently impoverished.

We must ensure that the true natural process can take place in the right way. What are called "stimulant effects" are actually the most important effects of all. It is just the substances that are regarded as unnecessary which are actively present in fine quantities all around the Earth, and plants need them just as much as they need what comes to them from the Earth. These substances are absorbed by the plants from the surrounding universe: mercury, arsenic, silicic acid. The plants absorb them out of the soil after the substances have been radiated into it. We human beings, however, are quite capable of preventing the soil from properly receiving what the plants need from the surrounding universe. If we continue to fertilize haphazardly, we may gradually prevent the soil from absorbing what is active in the finest homeopathic doses as silicic acid, lead, and mercury. What comes in from outside in this way is needed by the plants to help them build up their bodies in the form of carbon.

For this reason, we cannot stop with what I mentioned yesterday; the manure we use must be treated further. It is not a question of merely augmenting the manure with substances that we believe will be of benefit to the plants. It is a question of infusing the manure with living forces, which are much more important to the plants than the material forces, the mere substances. Although we might gradually make our soil especially rich in one substance or another, that would not help the plants unless our manuring also enabled them to absorb what the soil offered. That is the important thing. People today are totally unaware of the potent effect that tiny quantities can have on living things. However, Dr. Kolisko's[2] brilliant investigations into the effects of the smallest entities have placed what was blind groping into homeopathy onto a sound scientific basis. I believe this has made it scientifically acceptable to say that just those radiant forces needed by the organic world are released when small quantities of substance are applied in the appropriate way.

When it comes to manuring, it will not be at all difficult for us to use small quantities in the proper way. We have already seen how to make the cow horn preparations and how to apply them either before or after manuring. They supply forces that enhance the effectiveness of the manure that is used normally. We must

---

2. Lili Kolisko (1889-1976) was the director of the Biological Institute, a laboratory in Stuttgart, Germany.

also experiment with other ways of giving the manure the right degree of vitality and the appropriate consistency. We must enable it to retain of its own accord the proper amount of nitrogen and other substances that it needs in order to bring vitality to the soil. Today I would like to indicate some general guidelines for working with substances that can be added to the manure in small doses—in addition to what we take out of the cow horns—so that the manure can become vitalized and can transmit this vitality to the soil where the plants will grow. I am going to mention specific substances, but let me emphasize right from the beginning that if one or another of them is difficult to find in a given area, it is possible to replace it with certain others. There is only one case in which substitution is not possible, because what is characteristic of this particular plant is not likely to be found in any other species.

To begin with, we need to make sure that the major elements of the organic realm—carbon, hydrogen, oxygen, nitrogen, and sulfur—can come together in the right way with other substances, with potash salts, for instance. It is generally known that plants require a certain amount of potash salts for growth and that potash salts (or potash in general) tend to restrict the growth process to those parts of the plant that usually become the solid framework, that is, to the stems and stemlike components. But in the context of the interaction between the plant and the soil, it is important to transform this potash so that it relates properly to the forming of the proteinaceous material that constitutes the actual body of the plant. This can be achieved in the following way.

Take yarrow (*Achillea millefolium*), a plant that is usually easy to obtain. If it does not happen to grow in your area, you can use the dried herb just as well. Yarrow is actually a miracle of creation. If we compare yarrow to any other plant, we will be deeply touched by the particular wonder of it. You know how the spirit uses sulfur to moisten its fingers when it wants to carry substances—carbon, nitrogen, and so on—to their proper organic destinations. The way yarrow appears in nature, it is as if some plant designer had used an ideal model to bring sulfur into relationship with other plant substances. You might say that in yarrow, as in no other plant, the nature spirits reach the height of perfection in their use of sulfur. If yarrow is brought into the realm of biological activity in the proper way, its effect within the animal or human organism is to correct weakness of the astral body. If we know this, we can trace the essence of yarrow still further, throughout the process of plant growth in nature. Yarrow is a great asset when it grows wild in the country, along paths or at the edge of fields where grain, potatoes, or other crops are being grown. It should not be allowed to become a nuisance, but it is never actually harmful; under no circumstances should anyone try to eradicate it. Like some sympathetic people in human society who exert an influence simply through their presence and not through what they say, yarrow's mere presence in areas where it grows abundantly is extremely beneficial.

Take the same part of the yarrow that is used medicinally—the umbel-like inflorescences. If possible, it is best to pick the fresh inflorescences and then let them dry just a bit. They do not need to dry very much. If one cannot get fresh yarrow but only the dried herb, one should first try to moisten the inflorescences a little with juice squeezed from the leaves. Even with dried leaves, it is possible to obtain the necessary juice if one makes a decoction. We always stay within the living realm. After taking a handful or two of yarrow and compressing it somewhat, we take the bladder of a deer and enclose the yarrow substance in it as best we can and then tie it shut. We will have a fairly compact mass of yarrow inside the bladder. Hang it up in the sunniest spot available, and let it stay there for the summer. When autumn comes, take it down and put it in the ground—but not too deeply—and let it spend the winter there. We thus have enclosed yarrow flowers—which can have gone to seed a bit—for a year within the bladder of a deer and exposed them to various influences, partly above ground and partly below ground.

They will take on a very characteristic consistency during the course of the winter; this material, once it is in this form, can be kept for as long as one likes.

Remove this material from inside the bladder and add it to a manure pile, which can be as big as a house. One does not have to put a lot of effort into this—the radiating effect will work even if we mix the material into the manure only a little bit. The radiant energy is so strong that it will influence the entire quantity of manure or liquid manure or compost no matter how much it is divided up. (Materialists will believe us here, since they themselves talk about "radium"!) The effect of this material derived from yarrow is so enlivening and refreshing that simply by using this treated manure in the ordinary way, much will be done to counteract the unavoidable exploitation of the land that comes about through raising crops. In this way, the manure once again can enliven the soil so that the soil can absorb and retain the extremely fine doses of silicic acid and lead and so on that come toward the Earth. The members of the Agricultural Circle can experiment with this method; they will see that it really does work.

Because it is always preferable to work out of insight, the following question arises. We have become familiar with the yarrow plant: it can radiate its effects through large masses of manure because its own highly dilute sulfur content is combined with potassium in an ideal way. But why must we put the yarrow specifically into the bladder of a deer? Here we need to gain insight into the entire process related to this bladder. A deer is a creature that is intimately related not so much to the Earth as to the Earth's surroundings—to the cosmic aspect of the Earth's surroundings. That is why deer have antlers, which have the function I explained yesterday. What is present in yarrow is especially strongly preserved in the bodies of humans and animals by means of the process that takes place between the kidneys and the bladder. This process is dependent on the material constitution of the bladder. As thin as it may be in terms of substance, in its forces a deer bladder is almost a replica of the cosmos. A deer is involved with forces that are quite different from those with which a cow is concerned, which are all related to the interior of the body. By putting the yarrow into a deer bladder, we significantly enhance its inherent ability to combine sulfur with other substances. Handling yarrow in the way I have indicated is thus a fundamental means for improving manure—and one that always stays within the living realm. It is important to note that we never delve into inorganic chemistry; all that we do stays within the realm of the living.

Let us take another example. If we want to imbue manure with the possibility of absorbing vitality that it can impart to the soil where the plants are growing, we must make the manure especially able to combine the elements that are necessary for plant growth. In addition to potash, these elements include various compounds of calcium. In the case of yarrow, we are dealing primarily with the effects of potash; if we want to draw in the effects of calcium as well, we need another plant. This plant may not arouse our enthusiasm as readily as yarrow, but it, too, contains homeopathic sulfur, which enables it to attract the other substances it needs and incorporate them into an organic process. This plant is chamomile, *Chamomilla officinalis*.[3]

We cannot say simply that chamomile is characterized by a high content of potash and calcium. Rather, the situation is as follows: Yarrow develops its sulfur forces primarily in the potash forming process; it therefore contains sulfur in exactly the amount needed for working on potash. Chamomile, on the other hand, works on calcium in addition to potash and thereby develops what can help ward off the harmful effects of fructification and keep the plant healthy. The marvel is that, like yarrow, chamomile contains some sulfur, but in a different quantity because it has to work on calcium as well. The results of spiritual

---

3. *Chamomilla officinalis*, the German or true chamomile, is now classified as *Matricaria chamomilla*, or sometimes also as *Matricaria recutita*.

research are always derived from the greater whole, from macro-
cosmic rather than microcosmic relations. We can see that this is
true if we trace the process that chamomile goes through after
having been ingested by a human being or animal. Here the bladder is of no particular significance, but the substance of the walls
of the intestines is of great significance. If we want to work with
chamomile as we did with yarrow, we must pick its beautiful, delicate, white and yellow flower heads and treat them just as we did
the yarrow umbels; however, we must stuff them into cattle intestines rather than a bladder. Instead of using the intestines for
sausage casings, we simply make a different kind of sausage by
filling them with chamomile flowers. Notice how in this case,
too, we remain within the realm of the living.

All that remains to be done is to expose this material to natural influences in the right way. Since we want the chamomile to
be worked on by a vitality that is as closely related to the earthy
element as possible, we need to take these precious little
sausages—and they really are precious—and let them spend the
entire winter underground. They should be placed not too deeply
in soil that is as rich as possible in humus. We should also try to
choose a spot that will remain covered with snow for a long time
and where this snow will receive as much sunshine as possible, so
that the cosmic astral influences will work down into the soil
where the sausages are buried. Then, when spring comes, we can
dig up the sausages and store them or add their contents to
manure, just as we did with the yarrow. We will find that the
manure has not only a more stable nitrogen content than other
manures but also the ability to enliven the soil such that plant
growth is extraordinarily stimulated. Above all, we will raise
much healthier plants if we fertilize in this way.

To our modern way of thinking, this all sounds quite insane. I
am well aware of that, but just think of how many things were
originally rejected as crazy and after a few years became accepted.
I wish you could have read the Swiss newspapers when someone
first thought of laying train tracks up the mountains—you would

not believe how they ridiculed that poor person! But a short time later, the mountain railways were built, and it no longer occurs to anyone to say that whoever first thought of building them was crazy. It is just a question of overcoming people's prejudices. As I said, if these two particular plants are difficult to obtain, you can substitute another, although it is really better not to do so. In both cases, the dried herbs work just as well.

On the other hand, there is one plant whose beneficial influence on manure is such that it would be next to impossible to find a substitute for it. We are generally not very fond of this plant, at least not in the sense of wanting to fondle it, because the plant in question is stinging nettle (*Urtica dioica*). But stinging nettle is the greatest benefactor of plant growth, and it cannot be replaced by any other plant. If it is not available locally, the dried herb must be obtained from somewhere else. Stinging nettle is a jack of all trades; it has many different influences. It, too, contains sulfur, which, as I have already explained, plays an important role in assimilating and incorporating the spiritual element. Stinging nettle likewise carries the radiations and currents of potash and calcium, and it has a kind of iron radiation that is nearly as beneficial for the whole course of nature as the iron radiations in our blood are for us. Because it has such a beneficial influence, the stinging nettle we find growing in the wild does not deserve our customary scorn. It should actually be growing all around our hearts, since the role it plays in nature by virtue of its marvelous inner structure and way of working is very similar to that of the heart in the human organism. Stinging nettle is indeed a great boon.

If it should ever become necessary to reduce a soil's iron content—please excuse me, Count Keyserlingk, if I become too local with my example—planting nettles in out of the way spots can do a great deal toward freeing the upper layer of the soil from the influence of iron. Nettles like iron so much that they draw it out of the soil and into themselves; although this does not rid the soil of iron as such, it at least undermines the effect of iron on the growth of other plants. Planting stinging nettles around here, therefore, would be of particular importance. I only wanted to mention this in passing to make you aware that the mere presence of stinging nettle can already be of significance for the plant growth in its surroundings.

To improve the manure still more, gather stinging nettles and let them wilt slightly. Then compress them a bit and put them straight into the ground without any bladder or intestines. One can use a thin layer of loose peat or something similar to separate them from the surrounding soil. We must bury them right in the ground, marking the place carefully so that we do not dig out plain soil when we come back for them. Let the nettles spend the winter and also the following summer in the ground; they need to be buried for an entire year. Then we will have a substantiality that is extremely effective. When we add this to manure—just as we did with the other preparations—the effect will be to make the manure inwardly sensitive and receptive, so that it acts as if it were intelligent and does not allow decomposition to take place in the wrong way or let nitrogen escape. This addition makes not only the manure but also the soil more "intelligent," so that it individualizes and conforms to the particular plants that grow in it. The addition of stinging nettle in this form is like an infusion of intelligence for the soil. Modern methods of improving manure can sometimes produce astonishing outward results, but ultimately they tend to turn all first-rate agricultural products into mere stomach fillers. They will no longer have true nutritive value for human beings. It is important not to be deceived by produce that looks big and swollen; the important point is that the outward appearance be consistent with real nutritive content.

It can happen that plant diseases appear on our farms. I want to speak about them in rather general terms. People today like to specialize, so they talk about specific diseases. In the pursuit of science, that is all well and good, because we do need to know the characteristics of each disease. In the end, however, it is much the same as with a doctor: being able to describe an illness is usually not very useful; it is much more important to be able to cure it. Bringing about healing requires us to work from totally different points of view than we do when we are describing diseases. We can achieve the greatest degree of precision in describing illnesses and know exactly what is going on in the diseased organism according to the principles of modern physiology or physiological chemistry and yet still not be able to effect cures. We cannot heal according to histological or microscopic findings; in order to heal, we must understand the broad interrelationships.

That is how it is in the plant world too. Since plants are simpler organisms than animals or humans, healing can take place on a more general level. With plants one can use a kind of universal remedy. If this were not the case, we would be in a difficult position, much as we often already are when we treat animals. We are not in this difficult position with human beings, because people can tell us what is hurting them, whereas animals and plants cannot. A large number of plant diseases can be alleviated just by means of a rational method of manuring. It is necessary that manuring add calcium to the soil. For the calcium to have a healing effect, however, it has to be calcium from a living organism; we cannot evade the organic realm. It will not help at all to add ordinary lime or any other calcium compound that has fallen out of the organic realm.

The oak contains plenty of calcium. Seventy-seven percent of its substance consists of finely distributed calcium.[4] Oak bark, in particular, represents a kind of intermediate between the plant and the living earth element, in the same sense as I already described the kinship between bark and living earth. Of the many forms in which calcium can appear, the calcium structure of oak bark is ideal. Calcium has the effect that I have already described. It creates order when the etheric body is working too strongly, so

---

4. The figure of 77% is evidently based on an ash analysis of *Quercus robur* (pedunculate oak) which is indigenous to Eurasia and should not be confused with the American *Quercus rubra* (red oak).

that astrality cannot influence whatever organic entity is involved. Calcium in any form will kill off or dampen the etheric body and thereby free up the influence of the astral body. When we want rampant etheric development to uniformly and beautifully contract, without any shocks, we need to use calcium in the particular form in which it is found in oak bark.

For this purpose we collect whatever oak bark can be easily obtained; we do not need a great deal. Then we chop it up until it has a crumblike consistency, put it into a skull from any one of our domestic animals, and close up the skull, preferably with a piece of bone. Next we place the skull in a relatively shallow hole in the ground, cover it with loose peat, and set up some kind of pipe or gutter so that as much rainwater as possible flows into the hole. We might even put the skull into a rain barrel, where water can constantly flow in and out. Add some kind of plant matter that will decay, so that the oak bark in its bony container lies in this organic muck for the winter or, better still, for autumn and winter. Water from melting snow will do just as well as rainwater. When this material is added to a manure pile, it will truly provide the forces to prevent or arrest harmful plant diseases.

We have now added four different substances to the manure. All this takes a certain amount of work, but if we stop and think about it, it is less work and less money than all the fooling around in chemical laboratories that takes place in the name of agriculture. What we have discussed here is much more economical. We still need something else, something that will draw in the silicic acid from the cosmic surroundings. Plants need silicic acid, but it is precisely with respect to the absorption of silicic acid that the soil loses strength over time. It happens very gradually, so we do not notice it. People who look only at the microcosm and not at the macrocosm are not concerned about this loss of silicic acid, because they do not think it is significant for plant growth. In point of fact, it is of utmost importance.

What we are talking about now is less likely to be regarded by educated people as a sign of utter derangement than it would have been not too long ago. Today, the transmutation of elements is spoken about quite openly. Observation of many elements has tamed the lions of materialism in this instance. Still, certain things that are going on around us all the time are completely unknown. If they were known, it would be easier for people to accept such things as I have just been speaking about.

I am perfectly aware that people who are caught up in current ways of thinking will object that I have not said anything about how to improve the nitrogen content of manure. In fact, I have been talking about it all the time when I spoke of yarrow, chamomile, and nettle, because in organic processes there is a hidden alchemy at work. In this hidden alchemy, potash, if it is functioning properly in organic processes, can really be transformed into nitrogen. Even lime can be transformed into nitrogen. All four of the elements that I discussed have a role in plant growth. In addition to sulfur, there is also hydrogen. I have already talked about the significance of hydrogen. There exists between lime and hydrogen a mutual qualitative relationship that is similar to the qualitative relationship of oxygen and nitrogen in the air. Even a purely outward procedure, such as quantitative chemical analysis, can show that there is a relationship between the association of oxygen and nitrogen in the air and the association of lime and hydrogen in organic processes. Under the influence of hydrogen, lime and potash are constantly being transmuted, first into something resembling nitrogen and then into nitrogen itself. The nitrogen that comes about in this way is extremely useful for plant growth, but we must allow it to be produced by means of the methods I have described.

Silicic acid, as we know, contains silicon. In a living organism, silicon, too, is transmuted into an extremely important substance, one that is not currently included among the chemical elements. Silicic acid draws in the cosmic factor, and there must be a comprehensive interaction between silicic acid and potassium (not calcium) in the plant. We must enliven the soil through manuring so that it can facilitate this interaction. To this end, we need to look for a plant whose own potassium silicic acid relationship will allow it to impart this power to the manure, when it is added in a homeopathic dose. This readily available plant is the dandelion (*Taraxacum*). Just to have it growing on a farm is beneficial. The innocent yellow dandelion is a tremendous asset, because it mediates between the fine homeopathic distribution of silicic acid in the cosmos and the silicic acid that is used over the whole region. The dandelion is really a kind of messenger from heaven. If we want to make it effective in the manure, we have to utilize it in the right way. It must be exposed to the Earth's winter influence, but to acquire the surrounding cosmic forces, we must treat it just as we have the yarrow, chamomile, and nettle.

Collect the yellow heads of the dandelion and let them wilt a little. Then pack them together, sew them up in a bovine mesentery, and put them in the Earth through the winter.[5] When these balls are dug up in the spring, they will be thoroughly saturated with cosmic influences and can be stored until needed. This material can be added to the manure, and it will give the soil the ability to attract just as much silicic acid from the atmosphere and from the cosmos as is needed by the plants. In this way the plants will become sensitive to everything at work in their environment and will become able to draw in whatever else they need.

Even plants, in order to grow properly, need to have a certain ability to sense and perceive. In the same way that I can pass by a dull and insensitive person without being noticed, so can everything within and above the soil pass by a dull and insensitive plant. What the plant does not perceive, it cannot put to use for its own growth. But if a plant is very delicately permeated and

---

5. The abdominal cavity of mammals is lined with a membrane called the *peritoneum*. The stomach and intestines are suspended from the dorsal side of the abdomen by a sinuous fold in the peritoneum known as the *dorsal mesentery*. The portion of the mesentery supporting the stomach is itself folded and extended to form an apron over the front of the abdomen. It is not clear whether Steiner had a particular portion of the mesentery in mind.

enlivened by silicic acid, it becomes sensitive to all that is around it and can attract what it needs. It is very easy to weaken plants to the point that they can take advantage only of what is present in their immediate environment, which is not good. If we treat the soil as I have described, the plants will be able to draw in what they need from a very wide area. They will then be able to use not only what is in their own field but also what is in the soil of a nearby meadow, if they happen to need it. A plant can take advantage of what is in the soil of a neighboring forest, if it has been inwardly sensitized in this manner. Such an interplay can be brought about by imbuing the plants with the forces that the dandelion makes available to them.

We should try to produce fertilizer by enriching the manure with these five ingredients—or appropriate substitutes—in the ways I have suggested. Fertilizers of the future should not be prepared with all kinds of chemicals but rather with yarrow, chamomile, nettle, oak bark, and dandelion. A fertilizer of this kind will contain much that plants need. Even better, before using this treated manure, press the blossoms from the valerian plant (*Valeriana officinalis*) and greatly dilute the extract with warm water. The extraction can be done at any time and can then be stored. If this diluted valerian juice is applied to the manure in very fine doses, it will stimulate the manure to relate in the right way to the substance we call phosphorus. By using these six ingredients, we will be able to produce an excellent fertilizer, from liquid manure as well as from solid manure or compost.[6]

---

6. The six compost preparations described here are also known as Preparations 502 through 507.

DISCUSSION

(QUESTION) *When you talked about the deer bladder, were you referring to the male red deer, the stag?*
(DR. STEINER) Yes, I meant the male red deer.[7]

Did you mean the annual or the perennial stinging nettle?
Urtica dioica.

Should manure pits be roofed over in areas of heavy rainfall?
The manure ought be able to withstand any ordinary amount of rainfall. There is no hard and fast rule; it is just as bad for it to get no rain at all as it is for it to be leached out by rain. In general, rainwater is good for manure.

But should we not cover the places where manure is stored to prevent the liquid manure from being lost?
In a certain sense, manure needs rainwater. It might possibly be a good idea to keep some rain off by spreading peat over the manure, but it would not serve any purpose to roof the piles and keep the rain off altogether. The manure would certainly be the worse for it.

If plant growth is promoted by the methods of manuring you indicated, are cultivated plants and weeds affected equally, or do we need to use any special methods to get rid of the weeds?
That is a totally justified question, and I will be talking about weed control in the next few days. What I have mentioned so far

---

7. The red deer (*Cervus elaphus*) is native to Europe, North America, Africa, and parts of Asia. In North America the most closely related species is the wapiti or elk (*Cervus canadensis*), which is sometimes also considerd a subspecies of the red deer.

is conducive to plant growth in general, and it will not help get rid of weeds. These methods do make plants more resistant to parasites, so you thus have a means of controlling this kind of infestation. But since the general pattern of plant growth is also followed by weed growth, weed control has nothing to do with the principles we have discussed. We will talk about this subject later. These matters are so interconnected that it would be best not to discuss any one aspect separately.

What do you think about the Captain Krantz method of handling manure? It makes the manure odorless by relying on the warmth that is generated when the manure is piled in loose layers.

I deliberately refrained from saying anything about things that are already being done along rational lines. My intention was simply to show how the methods that arise from spiritual science can be used to improve ordinary methods. Although I am sure there are many advantages to the method you mention, I also believe it has not been in use very long, and it may turn out to be one of those things that are dazzling successes to begin with but prove to be not so practical in the long run. Initially, when the soil is still stuck in its traditions, so to speak, almost anything will give it a boost. But when you apply the same method over a longer period of time, it is similar to what happens with medicines when they first enter the body: the most unlikely remedies can help at first, but after a while the therapeutic effect wears off. In this case, too, it can take some time to realize that things are not working as you had hoped.

The spontaneous heating up is certainly significant, and there is no doubt that whatever you to do to make it happen is very good for the manure and will have positive results. However, it is possible that piling the manure loosely may not be good for it. I also wonder whether it is quite true that the manure becomes completely odorless. If it actually does become odorless, that would be a sign that you are doing something right. This method has not been in use for very many years yet, so it is too soon to tell.

Isn't it better to store the manure above ground than in pits?

In principle, it is good to pile the manure as high as possible, but the site itself should not be too high. It is important to maintain a proper relationship to the forces that are below ground. The manure should not be stored on a hillock, but you can certainly pile it up from the ground level, and that would be the best height.

Should we prepare compost in the same way if it is to be used on grapevines, which have suffered so much recently?

Yes, but with certain modifications that I will mention when we talk about orchards and vineyards. As a general rule, what I said today is valid for any kind of fertilizing, for improving all your fertilizing materials. We still need to discuss how to adapt this specifically for pastures and hayfields, grains, orchards, vineyards, and so on.

Is it all right if the storage area for the manure is cobbled or paved?

From everything we know about the structure of the soil and its relationship to manure, it would be a mistake to have your storage area paved. I cannot see any reason for it. The area underneath the manure and immediately around it should not be paved, so that the manure can interact with the soil. Why do damage to the manure by separating it from the soil?

Does it make any difference whether the ground underneath is sandy or clayey? Sometimes people cover the area with clay to make it impervious to water.

The different kinds of soil will certainly have different influences, which will depend on their particular characteristics. If you have sandy ground under your manure piles, you will have to mix in some clay to make it less porous before you pile up the manure. If, however, your soil is definitely on the clayey side, you will want to loosen it up by mixing in sand. For an intermediate effect, you can always take a layer of sand and then a layer of clay.

That way you will get both the earthy solidity and the watery influences; otherwise the water will just trickle away. A mixture of both kinds of soil is especially good. For the same reason, you should avoid, if at all possible, building your manure piles on loess soil or anything like that.[8] Such soils will not be very effective. In such cases it would be better to gradually build up a special base for the manure pile.

With regard to growing the plants you mentioned—yarrow, chamomile, nettle—is it possible to introduce them into the area, if they are not there already, simply by scattering the seeds? In our pasture management we have been assuming that both yarrow and dandelion were dangerous for cattle and have been trying to get rid of them along with the thistles. Would we now have to plant them again along the edges of the fields but not in the meadows and pastures?

Yes, but in what way are they supposed to be harmful for the animals?

(Count Keyserlingk) Yarrow is said to contain toxic substances, and dandelion is not supposed to be good for cattle.

You need to observe that carefully. In the open field the animals will not eat it.

(Count Lerchenfeld) In our area, it is just the opposite. Dandelion is considered good fodder for milking animals.

These are sometimes no more than opinions; nobody knows whether they have actually been put to the test. It is also possible that in hay these plants are not harmful. This would have to be tested. I think that if they were harmful, the animals would leave them alone, even in hay. Animals do not eat things that are bad for them.

---

8. Loess is a soft, porous rock that gives rise to rich, silty soils.

Hasn't yarrow been more or less driven out by heavy liming? Doesn't it need a damp, acidic soil?

If you use yarrow from the wild, a very small amount of it will be enough for a huge estate. Its pronounced homeopathic effect is what I had in mind. Having the yarrow in the garden here would be enough for the whole estate.

I have seen how cattle in my pastures eat young dandelions eagerly until they start to flower. Once they come into flower, the cattle will not touch them anymore.

You must remember, as a general rule, that animals will not eat dandelions if they are bad for them; animals have excellent feeding instincts. Another thing to remember is that when we want to promote a process, we ourselves often use something that we would not use all by itself. For example, no one would eat baker's yeast as a food, but we still use it in making bread. Something that may be poisonous in one set of circumstances, and if taken in large doses, might have a most desirable effect under other circumstances. After all, most medicines are poisonous. It is not the substance itself but how it is used that is important. Thus, I do not think you need to be concerned about dandelions being harmful to your animals. There are so many strange notions around. It is really quite curious that Count Keyserlingk emphasizes how harmful dandelions are, while, on the other hand, Count Lerchenfeld talks about them being the very best fodder for milking animals. The effect cannot be different in two such closely neighboring areas. One of these two opinions must not be right.

Is it perhaps the ground underneath that makes the difference? My statement was based on the opinions of veterinarians. Should we actually plant more yarrow and dandelion in our meadows and pastures?

A very small area will be enough.

Does it make any difference how long the preparations are kept in the manure after taking them out of the ground?

Once they have been mixed in with the manure, it does not matter how long they stay there. When you spread the manure on the fields, the mixing should already have been done.

Should the manure preparations be put into the ground together, or should each one be buried separately?

This is actually of some importance. For the sake of what is happening to them while they are in the ground, it is best if the preparations do not interfere with one another; for this reason, they should be buried some distance apart. If I had to do it on a small farm, I would choose places on the periphery, as far apart as possible. On a large estate, you can place them as far apart as you like.

Is it all right to have plants growing in the soil on top of where the preparations are buried?

That soil can do whatever it wants. In cases like this, it is quite good to have the area overgrown with plants, even with cultivated plants.

How should we put the preparations into the manure pile?

I would suggest the following procedure. I would put the preparations at least 10 inches into the top of a large pile of manure, so that they are surrounded by the manure. The manure should completely surround the preparations, then the rays—everything depends on the radiations—all go into the manure. It is not good if the preparation is too close to the surface. At the surface the radiation bends; it makes a distinct curve and does not leave the pile if the preparation is surrounded by manure. Twenty inches is deep enough, but if it is too close to the surface, a lot of the radiant force will be lost.

Is it enough to make just a few holes, or is it better to disperse the preparation as much as possible?

It is definitely better to disperse it and not make the holes just in one location, otherwise the radiations disturb each other.

Should we put all the preparations into the manure pile together?
When you put the preparations into the pile, you can put them in side by side. They do not influence each other, they only influence the manure as such.

Can we put all the preparations into a single hole?
We could assume on theoretical grounds that the preparations would not interfere with each other, even if you put all of them in a single hole, but I hesitate to state this beforehand as a definite fact. You can place them near one another, but if you were to combine them all in one hole, they might in fact interfere with each other.

What kind of oak did you mean?
Quercus robur.

Does the bark have to come from a live tree, or can it come from a felled tree?
From a live tree if at all possible, especially from one in which you can assume that the oak resin is still quite active.

Is the entire bark needed?
Only the surface, the outermost layer of bark, which crumbles when you remove it.

Is it absolutely necessary to stay within the topsoil when burying the manure preparations, or can the cow horns be buried deeper?
It is better to stay within the fertile layer. In fact, you can assume that if they are buried in subsoil, what you get out of them will not be as good. A deeper topsoil layer, of course, would be ideal. The spot with the deepest layer of topsoil will certainly be the best. You will not get any useful results below the topsoil.

If they are buried in topsoil, they will always be subject to frost. Won't that harm them?

The time when they are exposed to frost is precisely the time when, because of this frost, the Earth is most intensely exposed to cosmic influences.

How should we grind the quartz or silica? With a small mill or with a mortar and pestle?

It is best to do it in a mortar to begin with, and you will need an iron pestle. Grind it to a very fine, mealy consistency. If you use quartz, you will first need to grind it as far as possible with the mortar and pestle, and then grind it more on a glass surface. It has to be a very fine powder, which is very difficult to achieve with quartz.

Farming experience shows that well-fed cattle tend to put on fat, so there must be some connection between what they eat and what they absorb as nourishment from the atmosphere.

Consider carefully what I said: With the consumption of food, it is the forces developed in the body that are important. For an animal to develop sufficient forces to be able to absorb and assimilate substances coming from the atmosphere, it has to be eating the right food. Perhaps this comparison will help. If you need to put on a very tight glove, you cannot simply stuff your hand into it; you first have to stretch and enlarge the glove with a piece of wood or something. It is the same in nutrition: the inner forces that receive from the atmosphere what does not come from the food must first be made more supple. By means of food, the organism is stretched and enlarged and thereby enabled to absorb more from the atmosphere. This can even lead to hypertrophy, if too much is absorbed. You then pay for it with a shorter life span. There is a happy medium between too much and too little.

7

Agriculture III

THE FORCES ENVELOPING THE PLANT and supporting its growth are drawn to Earth from the cosmos and work into plants from below ground—through the soil—or are taken into plants from the surrounding air. The planetary influences that come down from Mercury, Venus, and the Moon and penetrate the soil promote the seasonal reproduction of plants, beginning and ending with seed formation. The forces streaming from Jupiter, Mars, and Saturn influence the fruiting of plants. Through the stimulus of the distant planets, plants burgeon and swell, while the near planets support plant propagation into successive seasons.

When it comes to addressing the question of how to combat weeds and various plant pests and diseases, we must consider the many-faceted planetary effects in the realm of agriculture, at various times of the year when the planets are aligned differently with respect to each other. Such pests as dandelions, field mice, and grubs can easily be eliminated from the garden. Burning the seed of the dandelion, the skin of the mouse, or the entire insect and spreading the ash over the garden plot or field will make an uncongenial home for these pests and discourage their return. Plants flourish under lunar water-saturated influences, which give life to the soil. If the soil becomes overly saturated, the life force intensifies in the upper portions of the plant and makes them prey for parasites and fungi. A diluted preparation of horsetail can combat this condition of the soil and eradicate these sorts of plant diseases. With insight into the vast, interconnected workings of nature on our planet and its cosmic influences and relationships, we can gain true knowledge concerning plant formation and growth and employ supportive agricultural methods.

LECTURE BY RUDOLF STEINER[1]
KOBERWITZ, JUNE 14, 1924

Over the past few days we have acquired insights into the growth and development of plants and also of animals, and these insights will provide a basis for our further considerations. We need to take at least a brief look at spiritual scientific ideas that relate to plants and animals as agricultural pests and to what are called plant diseases. In fact, very little can be said about these matters in general terms—they can be studied only in concrete detail. What I will do, therefore, is give examples, which may help serve as points of departure for experiments.

I would like to begin by dealing with weeds. For us, the definition of a weed is less important than the insight into how unwanted plants can be eliminated from specific areas. Indeed, do you know how people sometimes get strange impulses harkening back to their student days? In the grip of one of those impulses, I took a halfhearted look at a couple of textbooks, to see how weeds were characterized. Most authors say that a weed is anything that grows where we do not want it to grow. This sort of explanation does not bring us much closer to the essence of the matter. In

---

1. Lecture 6 from *(Agriculture) Spiritual Foundations for the Renewal of Agriculture.*

fact, we will never have much luck in finding the essence of a weed, simply because in the eyes of nature a weed has just as much right to grow as any plant we consider useful. We have to look at these things from a somewhat different point of view. The question for us is how we can keep a specific plant from growing in a certain area where it grows naturally under the prevailing conditions. We cannot answer this question without taking into account the things we have dealt with in the past few days.

We have seen the need to make strict distinctions among the forces active in plant growth. On the one hand, there are the forces that originate in the cosmos but which are first absorbed into the Earth and then work on plants. These forces—which essentially originate from the cosmic influences of Mercury, Venus, and the Moon but which work in a roundabout way via the Earth rather than directly from the planets—are the forces we need to take into account when we trace how one generation of plants leads to the next. On the other hand, in everything that plants acquire from the periphery, from what is above the Earth, we need to look to the various effects that the distant planets transmit to the air and which are taken up from the air. We can also say that all that comes from the near planets, in the way of forces working into the Earth, is heavily influenced by the lime in the Earth, while all that comes from the distant periphery is influenced by silica. Even when the influences of silica proceed from the Earth itself, they still transmit forces from Jupiter, Mars, and Saturn and not from the Moon, Mercury, and Venus.[2]

Today we are not in the habit of taking these matters into account, but we are also having to pay for it. In one recent instance, ignorance of cosmic influences—insofar as they work through the air and also as they are mediated by the Earth—had

---

2. The situation described here—the indirect influence of the near planets and the direct influence of the distant planets—had been alluded to in a previous lecture.

to be paid for in many parts of the civilized world. In these places, the methods that derived from old instinctive science had exhausted themselves. The soil was worn out, and so, too, were the agricultural traditions, even though sometimes the peasants were still able to help. What happened may not be a concern to you, but it affected many, many people. Far and wide, vineyards were infested with grape phylloxera, and there was practically nothing anyone could do about it. I still remember how the editors of an agricultural periodical in Vienna in the 1880s were bombarded with requests to find a means of combating the grape phylloxera and how they were at a loss as to what to do when the infestation became acute. These problems cannot be dealt with effectively by the science of today. The only effective way to deal with them is to delve into what can be learned along the lines we have indicated.

Let me show you this schematically (Steiner makes a blackboard drawing). Here is ground level, and what works in from the cosmos, from Venus, Mercury, and the Moon, then radiates back so that it works from below upward. The forces that work into the ground encourage plant growth of a single year and then culminate in seed formation. Out of the seed there grows a new plant and then a third plant and so on. Everything that emanates from the cosmos in this way flows into the plant's ability to reproduce, into the succession of generations.

On the other hand, everything that comes from above the surface of the Earth derives from different forces—those of the distant planets. Here is what transforms itself in the plant in such a way that the plant expands outward and becomes nice and plump. This is what we can remove for our nourishment, because there is a continuous stream building it up again. The parts we eat—the fleshy parts of apples or peaches, for example—all arise from the effects of the distant planets. Knowing this, we can understand what we have to do in order to influence plant growth in a particular direction. Taking these different forces into account is the only way to gain insight into how to affect the growth of plants.

A large number of plants—especially the ones we usually call weeds, even though some of the most powerful medicinal plants are found among them—are strongly influenced by what we can term "lunar effects." It is common knowledge that the surface of the Moon reflects the rays of the Sun, directing them back toward the Earth. We see these reflected rays of the Sun because we catch them with our eyes, and the Earth catches them too. The rays of the Moon are reflected Sun rays, but the Moon has imbued them with its own forces, and so they strike the Earth as lunar forces and have been doing so ever since the Moon separated from the Earth. This lunar force from the cosmos has an intensifying effect on all that is earthly. When the Moon was still united with the Earth, the Earth was indeed much more alive and fruitful. The Earth is more mineralized at present and is only barely strong enough to bring about growth in living things. Ever since the Moon separated itself from the Earth, however, it works to intensify the normal condition of the Earth, so that growth can be enhanced to take in reproduction.

When a living being grows, it gets big. The force at work in growth is also at work in reproduction. Simple growth, however, does not go far enough to produce a separate organism; instead, it merely brings forth cell after cell. Growth is a weaker form of reproduction, a type of reproduction that remains within the organism; reproduction itself is an enhanced form of growth. By itself, the Earth can impart growth, the weaker form of reproduction, but without the Moon it cannot produce enhanced growth. This requires the cosmic forces that shine down on the Earth from the Moon and, in the case of some plants, from Mercury and Venus as well.

As I said earlier, we usually imagine that the Moon simply takes up the rays of the Sun and reflects them onto the Earth. In other words, when we consider the effects of the Moon, we usually think only of sunlight, but that is not the only force that comes toward the Earth. Along with the moonbeams, the entire reflected cosmos emanates toward the Earth. The Moon reflects everything that comes toward it. In a certain sense, the whole starry heavens are reflected by the Moon and stream toward the Earth, though one could not prove it by any physical means available at present. It is indeed a very powerful cosmic organizing force that radiates from the Moon into the plants, so that the plants are also enabled to form seeds; in this way, the force of growth can be enhanced to become the force of reproduction.

For a given location on Earth, this force is available only at full moon. At the time of the new moon, that same area does not receive the benefit of lunar influences. The influences plants receive at the full moon will persist, but that is all. We could achieve significant results if at planting time, for instance, we were to utilize the Moon to support early germination—if we were to sow according to the phases of the Moon, as people in India did well into the nineteenth century. Still, nature is not so cruel as to punish us for the slightly inattentive and discourteous way we treat the Moon with regard to sowing and harvesting. We have the full moon twelve times a year; this suffices to ensure that there are always enough full moon effects, that is, enough of the forces that promote fruit formation. If we undertake fertility measures at the time of the new moon instead of at full moon, they simply lie dormant in the Earth until the next full moon, disregarding our human error and taking their cue from nature. Thus, people use the Moon without knowing it, but they do not go further.

Under such circumstances, weeds demand their rights just as do cultivated plants. We become confused because we are unacquainted with the forces that regulate growth. We must enter into them. Once we do, we realize that the fully developed lunar forces work for the reproduction of every kind of plant life. In other words, they encourage the forces that shoot up from the root into seed formation. We would get the best stand of weeds if we were simply to let the beneficent Moon have its effect on them and did not hinder it in any way. Weeds would reproduce and multiply, especially in wet years when the lunar forces work best. When we take these cosmic forces into account, we realize the following: If we prevent the Moon from exerting its full influence on weeds, and if we allow them instead to be influenced only externally by direct, nonlunar influences, their proliferation will be curbed. They will not be able to reproduce. Since we cannot simply switch off the Moon, we need to treat the soil so that it will become unsuitable for absorbing lunar influences. The weeds can acquire a reluctance to grow in soil that has been treated in a certain way. If we can accomplish that, we will have what we want.

If we see that weeds are making headway in a given year, we must simply accept the fact and calmly resolve to take action. We must gather a number of seeds from these weeds, since the force I have been talking about is enclosed in the seeds. We then light a fire—a simple wood fire is best—and burn the seeds and carefully collect whatever ashes are left. We will not obtain very much ash in this manner, but concentrated in the ashes of these seeds we have the opposite force to that which is developed in the attraction of the lunar forces. We take this little preparation that we have made from the various weeds and scatter it over our field. This ash has a large radius of influence, so we do not even need to be terribly careful in distributing it. By the second year we will notice that there are far fewer of the kinds of weeds we treated in this way and that they are no longer growing as well as they were. Because there is a cycle of four years with many plants in nature, we will find that after the fourth year the weeds that we have treated by scattering this pepper each year will no longer be present in this field.

Here we have an example of the fruitful application of the effects of the smallest entities, which has been demonstrated scientifically at our Biological Institute. A great deal can be achieved in this way. If one takes into account cosmic influences, which are ignored today, one can gain control over any number of things. For instance, we can feel quite free to plant as much dandelion as we need for the purposes I mentioned yesterday, but we can also burn its seeds and scatter the pepper over the fields. We would be able to have dandelions growing exactly where we want them, yet keep the treated fields free of them. Although people will not believe it, methods like this used to be understood and mastered out of an inborn agricultural wisdom. In earlier days, with these instinctive methods, people could plant everything together in quite limited areas.

I am simply giving indications to serve as starting points for practical application. Because people these days are of the opinion—I do not want to call it a prejudice—that everything has to be verifiable, we should try to verify our methods. If we conduct the experiments properly, the results will surely be confirmed. If I had a farm myself, I would not wait for the confirmation; I would start using these methods right away, because I am quite sure they work. As far as I am concerned, spiritual scientific truths are true in and of themselves and do not need to be confirmed by experiments. Our scientists have all made the mistake of wanting to verify spiritual scientific truths by external methods. This has happened even within the Anthroposophical Society, where people should know that things can be inherently true. But in order to make progress in this day and age, we have to verify these truths outwardly; we have to compromise. In principle, though, such compromise is not necessary.

How does one know inwardly that something is true? One knows it by virtue of its inner quality; one knows it for certain. We know it as certainly as we know, for instance, that if it takes fifty people to manufacture an item and we want to produce three times as many, we need a hundred and fifty people to do it. Of course, some smart aleck can come along and say, "I don't believe that a hundred and fifty people will make three times as many items, we will first have to put that to the test." It can then happen, under certain circumstances, that we are proved wrong by experience. Suppose we let the item in question be produced first by one person, then by two, and then by three, and suppose we determine statistically how many items the three people have produced. If the three people spent all of their time talking to one another, they will each have done less than one person. In that case, the assumption we made will prove false. The experiment can yield a contrary result, but that establishes nothing. If we proceed carefully, we must also carefully examine this contrary result. If we do, then what is inwardly true will also be confirmed outwardly.

We have spoken in quite general terms about weeds, but matters will not be as straightforward when it comes to animal pests. Let me choose an example that is particularly characteristic, so that experiments can be conducted to show how these methods are confirmed in practice. Let us take the farmer's best friend, the field mouse. What great lengths people go to trying to get rid of field mice! We can read in agricultural books that people have tried all kinds of phosphorus preparations, for instance, or a combination of strychnine and saccharin. Even more drastically, it has been suggested that field mice can be infected with typhus, by setting out a bait of mashed potatoes inoculated with certain bacilli harmful only to rodents. Such things have been done, or at least recommended. People have employed all sorts of rather inhumane methods of doing in these innocent looking little animals. I think even the government has had to intervene, because if one person is using one of these methods, it does no good unless his neighbor uses it too—the mice will just come over from the field next door. The government has to be called in to help so that everyone is forced to get rid of mice in the same way. The government is not interested in modifying its regulations; regardless of whether the method chosen is right or wrong, everybody has to go along with it. All must be done uniformly. Uniformity, after all, is the government's ideal.

All of this experimentation and regulation are very superficial. One always has the feeling that the experimenters themselves are not quite happy about it, because in the end the mice always come back. In our case, too, we need to use a method that is not restricted to a single farm. This method will help for a single farm, but it will not be entirely successful unless we can share our insight and persuade our neighbors to do the same. I am convinced that in the future we will have to rely much more on shared insights than on regulations enforced by the police. That will be a real step forward for society.

First, catch a fairly young mouse and skin it. There are always plenty of mice around, but for this experiment it must be a field mouse. It is important to obtain this mouse skin in the period when Venus is in the sign of the Scorpion.[3] Those old time farmers with their intuitive science were not so foolish after all, because as soon as we leave the plant kingdom and turn to the animals, we have to do with the zodiac, which means "animal circle." When we are trying to accomplish something in the plant world, we can stay within the solar system, but that is not enough when it comes to animals. In the case of animals, we need concepts that take into account the fixed stars, especially the fixed stars of the zodiac. With plants, the effect of the Moon by itself is just strong enough to push the growth process into reproduction, but in the animal kingdom the Moon's effect has to be supported by Venus. With animals we do not need to give much consideration to the effects of the Moon, because they preserve the lunar forces within themselves and emancipate themselves from the Moon. In other words, the forces of the Moon are developed and available in the animal kingdom even when it is not the time of the full moon; the full moon force in the animal is liberated with respect to time. Such is not the case with the other planetary forces, and this is what concerns us now.

We must burn the skin of the mouse when Venus is in the sign of the Scorpion and carefully collect the ash and anything else left over. There will not be much, but if we catch a number of mice, we will obtain enough. What is destroyed through fire at the time when Venus is in the Scorpion contains the negative of the field

---

3. Although Steiner refers here to the "sign" of the Scorpion (*Zeichen des Skorpions*), he is apparently not refering to the traditional astrological sign; in the discussion following this lecture, Steiner refers explicitly to the visible *constellation* of the Scorpion (*Sternbild Skorpions*).

mouse's reproductive force. We take the pepper we have made in this way and scatter it on our fields. If this is difficult to do in certain cases, we can use a homeopathic amount of the pepper—we do not need a whole plateful. If the pepper has been properly burned when Venus and the Scorpion are in peak conjunction, then in this pepper we will have a potent means of making the mice avoid the fields where it has been strewn. These animals have a lot of nerve; they will come right back and settle down in any unpeppered areas. Although the effect radiates quite far, it can happen that one is not quite thorough enough. However, if everyone in the neighborhood does this, the results will certainly be dramatic. I venture to say that we might even develop a taste for doing such things and come to enjoy them. Farms, like certain foods, taste better with a little pepper!

It is important to become accustomed to reckoning with the effects of the stars without being the least bit superstitious. A lot of things that used to be a matter of knowledge have degenerated into superstition, and it does no good to warm them up again. We have to start over again from knowledge, and it must be knowledge acquired in a spiritual way and not just by means of the physical senses. I have just described how we can treat the soil to combat those field pests that can be categorized among the higher animals. Mice are rodents, so they are considered higher animals. This method will have no effect against insects, because insects and other lower animals are subject to an entirely different set of cosmic influences.

I am going to tread on thin ice now and use the sugar beet nematode as an example, in order to speak about a current and local topic. The first signs of the presence of sugar-beet nematodes are the well-known swellings of the rootlets of the plant and also the fact that the leaves remain wilted in the morning. That is the outer indication. We must be clear that this leafy middle region—the part that undergoes a change in this case—absorbs cosmic influences from the air but that the roots absorb the forces that enter the plant from the cosmos by way of the soil. When nematodes appear, the process of absorbing cosmic forces, which should otherwise be taking place in the leafy part of the plant, is forced downward into the region of the roots Then, the cosmic forces that ought to be working up above are actually working below. The real phenomenon is that certain cosmic forces work too far down, and this is what causes the change in the plant's outward appearance. The cosmic forces that nematodes need for their survival become available in the ground, where the nematodes live. These little threadlike worms would otherwise have to live up in the leaves, but since the soil is their natural habitat, they cannot do that.

Certain living things—actually, all living things—can live only within a certain range of conditions. Just try living in air that is minus 70°C. We will find that humans are entirely dependent on a certain temperature range; above or below this range, we can no longer survive. Nematodes have similar limitations. They can live only in the soil and only if cosmic forces are present there at the same time. Otherwise they die out. Each living thing requires quite specific conditions for its survival. The human race would die out, too, if the right conditions were not prevalent.

For creatures that live in the particular way that nematodes do, it is crucial that cosmic influences, which should otherwise be active only in the surroundings of the Earth, enter into the Earth. These cosmic influences have a four-year cycle. With nematodes we are dealing with a phenomenon that is quite abnormal; this phenomenon we can also study by looking at cockchafer grubs, which emerge every four years. The same forces are at play. The forces that give the Earth the capacity to grow potato sprouts are the very ones that the Earth acquires for aiding the development of cockchafer grubs, which appear among potatoes every four years. Although this four-year cycle is not apparent with nematodes, it does play a part in what one has to do to counteract them.

We must take the whole insect and not just part of it, as with the mouse. A harmful root-dwelling insect such as the nematode is, in its entirety, a result of cosmic influences; it needs the Earth only as a substratum. We must burn the entire insect. Burning it is the best and fastest method. We could also let it decay; it is difficult to collect the end products of decomposition, though in some ways they might be better. In any event, we will certainly accomplish our objective by burning the whole insect. We may need to dry and store the insects, however, since the burning must be done when the Sun is in the sign of the Bull, which is exactly opposite the position of Venus when we make the mouse skin pepper. The whole insect world is related to the forces that develop as the Sun moves through the Waterman, Fishes, Ram, Twins, and on into the Crab; by the time the Sun gets to the Crab, these forces are quite weak, as they also are when it is passing in front of the Waterman. While the Sun is moving through this part of the heavens, it is radiating forces that have to do with the insect world.

People have no idea that the Sun is actually very specialized and that it is not at all the same entity when, in the course of the year or the day, it shines down on the Earth from the Bull or from the Crab and so on. It is constantly changing. In fact, it is rather absurd to speak of the Sun in general. We ought actually to say, "Ram Sun," "Bull Sun," "Crab Sun," "Lion Sun," and so forth. In each case the Sun is a quite different being, whose nature depends on the combined effects of its daily as well as its yearly course, which are determined by its position in the vernal point.

If we make an insect pepper, we can scatter it over fields of root crops, and the nematodes will gradually become powerless. After the fourth year, we will find that they have become quite helpless. They cannot survive; they shy away from life if they have to live in soil that has been peppered in this fashion. In a remarkable way, we have here what in earlier times was known as star knowledge. The knowledge of the stars that we have today serves only as a means of mathematical orientation. It is no longer of much use for anything else, but that was not always the case. In ancient times, people saw the stars as a guide for their earthly life and work. That science has now been lost.

With these methods we can keep harmful animals and insects at bay. It is important to develop a relationship to the Earth that allows for us to understand, on the one hand, that the Earth is enabled by the lunar and watery influences to bring forth plants. On the other hand, however, the plant and, indeed, every living thing also carry in themselves the germ for their own destruction. Just as water is indispensable for fertility, so is fire a destroyer of fertility. Fire consumes fertility. If something that is ordinarily treated with water to promote fertility is instead treated with fire, then within the household of nature we bring about the opposite of fertility, namely, destruction. That has to be taken into account. Under the influence of Moon saturated water, a seed develops fertility and proliferates; under the influence of Moon saturated fire—or fire saturated with any other cosmic force—the same seed spreads a force of destruction. It is really not so hard to believe that we reckon here with strongly spreading forces, if we also take the precise effects of time into account. The force in the seed spreads outward, and this works on in the force of destruction. Just as the seed has an inherent tendency to spread outward, so does the pepper that we produce from it. I call it "pepper" simply because it resembles pepper.

We still need to take a look at the so-called plant diseases. Actually, they should not be termed plant diseases, because abnormal processes in plants are not diseases in the same sense as they are in animals. (We will understand this difference better when we come to talk about the animal kingdom.) Most of all, plant diseases are not the same as disease processes in human beings. Actual illness is not possible without the presence of an astral body, which in an animal or a human being is connected with the physical body by means of the etheric body. There is a certain standard of normalcy for this connection, and most illnesses come about when the astral body unites unusually intensely with the physical body or some particular physical organ; that is, when the etheric body does not provide enough padding, so that the astral body pushes its way into the physical body more strongly than usual. Since a plant does not have an actual astral body, the particular way in which animals and humans become ill simply does not apply to plants. We must be quite aware of this fact and try to understand what it is that can cause plants to become ill.

You will have gathered from what I have said that the soil around plants has a certain inherent vitality. Accompanying this vitality are all kinds of growth forces, faint suggestions of reproductive forces, and also all the forces working in this direction under the water-mediated influence of the full moon. Although these forces in the soil are not so intense as to appear as actual plant forms, they are present with a certain intensity all around plants. There are several significant relationships: First we have the soil and also the soil saturated with water. Then we have the Moon. By letting its rays stream into the soil, the Moon makes the soil inherently alive to a certain extent; it induces waves in the etheric life of the soil. When the soil is saturated with water, it is easier for the Moon to have this influence; when the soil is dry, it is more difficult. The water is only the mediator. To be sure, the solid earth must be enlivened, that is, the mineral earth. But water, after all, has mineral qualities. A sharp boundary does not exist here.

It can easily happen, however, that the lunar influences in the soil become too strong. Suppose we have a very wet winter followed by an equally wet spring. The lunar force will then penetrate too strongly into the earthy element, and the soil will become too strongly enlivened. (Steiner made another blackboard drawing.) If the soil had not been too enlivened by the Moon, the plants growing aboveground would develop normally, up to and including the process of going to seed. Imagine wheat, or a similar plant, growing and growing until it eventually forms seeds. If the Moon imparts the right degree of vitality to the soil, this vitality works upward so that these seeds come into existence.

Let us assume, however, that the Moon's influence is too strong, that the soil is overly enlivened. In this case, the vitality works up too strongly from below, and something that should take place only in seed formation starts to happen earlier. When the vitality is too strong, it does not reach the top; its very intensity makes it start working lower down in the ground. Thus, through the effect of the Moon, there is insufficient force for seed formation. The seed incorporates a kind of dying life, and through this dying life a second ground level is formed above the level of the soil. Although there is no actual soil there, the same influences are present. As a result, the seed, or the upper part of the plant, becomes a soil for other organisms. Parasites and fungi appear—blights and mildew and the like. The forces that want to work upward out of the soil are kept from reaching the right height because of the overly strong lunar force. It is remarkable that this happens when the lunar forces are too strong rather than too weak, but so it is. A healthy seed-forming capacity is absolutely dependent on normal lunar forces, rather than lunar forces that are too strong. Theorizing and speculation, rather than perception, might lead to the opposite conclusion, but that would be wrong. Direct perception reveals what I have just described.

We need to relieve the soil of the excessive lunar force; we must find a way to reduce the water's mediating capacity, to give the soil more earthiness so that the water that is present does not absorb the excess lunar influence. We accomplish this by making a fairly concentrated tea out of *Equisetum arvense* [horsetail], which we then dilute and use as a liquid manure on the fields where we want to combat blight and similar plant diseases. Once again, very small amounts—a homeopathic application—will be sufficient.

This is an instance in which one can clearly see how the different fields of practical endeavor should interact. When we understand the remarkable effect of *Equisetum arvense* on the human organism through the functioning of the kidney, we will have a guideline for working. One cannot speculate about these things, of course, but this gives us a guideline for testing the way in which *Equisetum* works when it is made into a liquid manure.

When we spray it on the fields, even just a bit, it will work over wide areas. One does not need any special apparatus. *Equisetum* is an excellent remedy, although "remedy" is not quite the right word, since plants cannot truly become sick. It is not really a healing process; it is simply the opposite of the process I described earlier.

In this way, if we acquire insight into the various aspects of nature's workings, it is entirely possible to take hold of the processes of growth—not only plant growth but also the normal and abnormal growth of animals. Only at this point does real science begin. The type of experimentation taking place today is not true science; it is merely a recording of individual phenomena and isolated facts. Real science begins only when we are able to take hold of the effective forces.

The plants and animals on Earth, and even the parasites of the plants, cannot be understood in isolation. It is nonsense to seek within the compass needle itself for the reason why it always points to the north. That is not done; the whole Earth must be taken into account. When a magnetic north pole and magnetic south pole are assigned, the whole Earth is taken into consideration. But just as we have to look at the whole Earth when we want to explain how a compass needle behaves, so must we also consult the entire universe when it comes to understanding plants. It is not enough to look only at the plant, animal, or human kingdom. Life comes from the whole universe, not merely from what the Earth itself provides. Nature is a unity, with forces working in from all sides. Those whose eyes are open to these manifest forces will be able to understand nature. Today's science takes a little glass plate, puts on it a carefully prepared specimen, and peers at it through a microscope. That is the exact opposite of what we ought to be doing if we want to comprehend the full dimensions of the world. It is bad enough to be shut up in a room in isolation, but now we shut out the whole glorious world. Nothing remains but what we see through the lens of the microscope. This is what we have gradually come to. When we are able to find our way back to the macrocosm, then we will once again begin to understand something about nature and about many other things as well.

———

Discussion

(QUESTION) *Can the method you suggested for nematodes also be applied to other insects, to any kind of pest? Is it really permissible to use these methods to destroy plant and animal life over large areas without any misgivings? Serious abuses could arise. Surely some kind of limit needs to be set so that someone cannot just spread destruction all over the globe.*

(DR. STEINER) Just imagine if this method were not permissible. (I want to leave aside for a moment the ethical question, the issue of occult ethics.) What would happen in that case is that agriculture in the civilized regions of the world would deteriorate more and more, and near famine and high prices would cease to be isolated local phenomena and become the general rule. This will happen in the not too distant future, so our only options are either to let civilization go to ruin or to try to do things in such a way that a new fertility can come about. Faced with this need, we do not really have the option of debating whether these methods are permissible.

Nevertheless, from a different point of view, this question is quite valid. It is important to think about how to create a safety valve against abuse. If these methods become common knowledge, they could be abused. Still, I would like to point out that there were long epochs of civilization during which methods like this were known and very widely practiced, and yet it was possible to restrict access to responsible sectors of humanity so that they were not abused. Serious misuses did occur back at a time when much more severe misuses were possible, owing to the fact that the forces in question were active more widely. That was the case during the later periods of the Atlantean evolution, where serious abuse did indeed take place, leading to tremendous catastrophes.

In a general sense, I can only say that it is certainly justifiable to limit knowledge of these matters to a restricted circle, not to let it become generally available. In this day and age, however, this is almost impossible. Knowledge cannot be kept within small groups; it invariably gets out in one way or another. Before the invention of the printing press, it was easier, and it was easier still when most people did not know how to write. But now, whenever one gives a lecture, no matter how small the audience, the issue of where to find a stenographer always comes up. Indeed, I never like to see stenographers. I have had to get used to them, but it would be better if they were not present—I mean as stenographers, of course, not as people.

On the other hand, do we not also have to reckon with the necessity of moral improvement in all aspects of human life? That would be the panacea. Admittedly, if you look at certain contemporary phenomena, there is reason to be pessimistic; with regard to moral improvement, however, this pessimism should never lead us to become mere spectators. We should have thoughts that are filled with impulses of will, and we should always be looking for what we can do to raise the general level of human morality. This, too, could flow from anthroposophy, and an agricultural circle such as we have discussed could certainly act as a kind of remedy against such abuses.

In nature, too, it is certainly the case that what is beneficial can become harmful. If we did not have the lunar forces down there, we also could not have them up here. These forces must be active here; it is just that something that is eminently desirable and necessary in one place may be harmful somewhere else. What is moral at one level may be decidedly immoral on another. The Ahrimanic influence in the sphere of the Earth is harmful only because of where it is; when it works in a sphere that is only slightly higher, its effect is definitely beneficial. But to answer your first question, you are right. The method I indicated for nematodes applies to insects in general, to all the lower animals, which are largely characterized by a ventral nerve cord instead of a spinal cord. Where there is a spinal cord, you need to skin the animal. Where there is a ventral cord, you need to burn the whole animal.

Did you mean the wild chamomile?

This chamomile, the one that has petals turned down. It has petals directed downward rather than upward. It is the *Chamomilla officinalis* that grows wild by the roadside.[4]

Should we use the flowers of the stinging nettle?

Yes, but take the leaves too. Take the whole plant when it is in flower, but not the roots.

Can we use the dog chamomile (Hundskamille) *that grows in the fields?*

The dog chamomile is closer to the right type than the garden chamomile that is being passed around. Do not use the garden chamomile. The one you refer to, which is also used for chamomile tea, is much more closely related than this one here. You could use that one.[5]

The chamomile that grows along the railroad track here is the one you mean, isn't it?

Yes, that is the right one.

Does what you said about destroying weeds also apply to aquatic weeds, such as waterweed [Elodea canadensis]*?*

---

4. Steiner refers here to German chamomile (cf. lecture 5, note 3, p.143). This species is most reliably distinguished from several similar species by the hollow space in the center of its flower heads.
5. The "garden chamomile" that Steiner did not recommend using may have been *Chamaemelum nobile* (noble or roman chamomile) or *Chrysanthemum parhenium* (feverfew), both of which are commonly cultivated.

It also applies to weeds that come from the swamp or the water and to aquatic weeds. In this case you simply sprinkle the pepper on the banks.

Can the methods used against aboveground parasites also be used against underground parasites—cabbage clubroot for instance?
Absolutely.

Can these methods for alleviating plant diseases be applied to vineyards too?
I can only say that I am convinced that the vineyards could have been protected if people had gone about it in the way I have indicated. It has not been tested yet, however; little occult research has been done in this area or by me.

What about downy mildew?
That can be treated just like any other blight.

Is it actually right for us as anthroposophists to resuscitate the production of grapes for wine?
In many instances today all anthroposophy can do is point out the truth of the matter. The question of how things *ought* to be is a difficult one. I once had a good anthroposophical friend who kept extensive vineyards but who used a sizable share of the annual profits from them to send out postcards all over the world, advocating abstinence. On the other hand, I also had a friend who was a strict teetotaler and who was very generous to the anthroposophical movement throughout his life but who was also responsible for all those posters on the streetcars advertising "Sternberger Cabinett" [a fancy wine]! That is where the practical questions get tricky. Certain things are not possible in this day. That is why I said we should certainly use cow horns but that to become bull headed in our opposition to various things could be very harmful to the cause of anthroposophy.

Couldn't something else be substituted for stag bladder?

It is true that stag bladders may be hard to obtain, but a lot of difficult things get done in this world! Of course, you could experiment and see whether the stag bladder could be replaced by something else. I cannot say at the moment. It is entirely possible that somewhere there is another suitable kind of animal—perhaps indigenous to a corner of Australia. Among the animals native to Europe, I cannot think of anything else. One cannot use anything other than an animal bladder. I would not recommend that you immediately start looking for substitutes.

Is the constellation always the same when it comes to getting rid of insects?

That is something that has to be tested. As I said, the whole sequence from the Waterman through the Crab would need to be considered, and the constellations may vary significantly among the different lower animals. You will have to experiment.

With regard to getting rid of field mice, did you mean Venus in the astronomical sense?

Yes, the one we call the evening star.

What did you mean by "when Venus is in the Scorpion"?

That means any time when Venus is visible in the sky with the constellation of the Scorpion in the background. Venus has to be behind the Sun.[6]

---

6. It is likely that the preposition "behind" refers here to *time*, since neither Venus nor its fixed star background are visible when Venus is behind the Sun in terms of *space*. Thus, in terms of their daily westward motion, Venus is *behind* the Sun whenever it sets *after* the Sun. Venus is then an evening star, which corresponds with Steiner's reply to the previous question. Although Venus passes in front of the constellation of the Scorpion approximately once a year, only in four out of every eight years is Venus also an evening star at that time. (After eight years, Venus repeats its position almost exactly.)

When the upper parts of potato plants are burned, does it affect the growth of the potatoes?

The effect is so slight that you can disregard it. There is always an effect of some kind whenever you do anything with organic remains—and it affects the entire field, not just the individual plants—but in this case it is so slight that for practical purposes it can be disregarded.

What did you mean by "mesentery"?

I mean the peritoneum. To my knowledge, the mesentery means the peritoneum.

Is that the same thing as tripe?

No, it is not. It is the peritoneum.

How should we distribute the ashes over the fields?

You can take what I said quite literally: it is just like sprinkling pepper. It has such a wide radius of activity that it is enough if you simply walk over the fields and sprinkle it around.

Do the preparations work on fruit trees in the same way?

In general, everything I have said up to now may also be applied to fruit crops. There are a few other things still to consider, but I will tell you about them tomorrow.

In farming we usually apply barn manure to root crops. Is the prepared manure the right thing for grains also, or do they require some special treatment?

For the time being, you can go on manuring as usual; it is only a matter of supplementing with the methods I have described. With regard to matters I have not touched on, do not immediately assume the ordinary methods are all bad and that you have to completely alter them. Methods that have stood the test of time should be continued, supplementing with what I have indicated. The effect of the methods I suggest will be considerably modified, however, if you use a lot of sheep or pig manure. The effect will be much less striking if you use too much sheep and pig manure.

How is it when inorganic fertilizers are used?

People will discover that using inorganic fertilizers must eventually stop altogether. Any mineral fertilizer gradually reduces the nutritive value of whatever is grown in the fields where it is used; that is a general law. If you carry out just the measures I have indicated, however, they will make manuring unnecessary more than once every three years. Perhaps you will need to do it every four to six years. You will be able to do without artificial fertilizers altogether, if only because they will become an unnecessary expense. Artificial fertilizer will not be needed anymore; in time it will disappear.

Nowadays we take into account much too short spans of time. In a recent discussion on beekeeping, for instance, an up-to-date beekeeper came out in favor of breeding queens industrially, of selling the queens and distributing them widely, rather than having the individual beekeepers raise them. I had to say: "Of course, you are correct! But if not in thirty, then certainly in forty or fifty years you will find that this has ruined beekeeping."[7] Everything is being mechanized and mineralized today, but the fact is that what is mineral should work only in the way it does in nature. Unless you incorporate the mineral component into something else, you should not introduce minerals—or anything totally lifeless—into the living soil. This will not be feasible by tomorrow, but by the day after tomorrow it will certainly become a matter of course.

How should we catch the insects? Could they be used in the larval stage?

---

7. See *Bees*, Rudolf Steiner's nine lectures on bees published by Anthroposophic Press, 1998.

You can use both larvae and adults, but the constellation may change. As you go from the winged stage to the larva, you will need to move from Waterman toward the Crab. The right constellation for mature insects will be closer to the Waterman.

Further Reading

Adams, George. *Physical and Ethereal Spaces*. London: Rudolf Steiner Press, 1978.

_____ and Olive Whicher. *The Plant between Sun and Earth*. 2nd ed. London: Rudolf Steiner Press, 1980.

Anderson, Adrian. *Living a Spiritual Year: Seasonal Festivals in Northern and Southern Hemispheres*. Hudson, NY: Anthroposophic Press, 1993.

Baker, C.T.G. *Understanding the Honey Bee: Notes on the Organic Nature of the Bee Colony, and its Relation to Plant and Cosmos*. London: Anthroposophical Agricultural Foundation, 1948.

Bockemühl, Jochen, ed. *In Partnership with Nature*. Wyoming, RI: Biodynamic Literature, 1981.

_____. *Toward a Phenomenology of the Etheric World*. Hudson, NY: Anthroposophic Press, 1992.

Bortoft, Henri. *The Wholeness of Nature: Goethe's Way Toward a Science of Conscience Participation in Nature*. Hudson, NY: Lindisfarne Press, 1996.

Cloos, Walther. *The Living Earth: The Organic Origin of Rocks and Minerals*. Launceston: Lanthorn Press, 1996.

Cook, Wendy E. *Foodwise — Understanding What We Eat and How it Affects Us: The Story of Human Nutrition*. Forest Row: Clairview Books, 2003.

Edwards, Lawrence. *The Vortex of Life: Nature's Patterns in Space and Time*. Edinburgh: Floris Books, 1993.

Fyfe, Agnes. *Moon and Plant: Capillary Dynamic Studies*. Arlesheim, Switzerland: Society for Cancer Research, 1975.

Goethe, Johann Wolfgang. *Goethe's Botanical Writings*. Trans. by Bertha Mueller. Woodbridge, CT: Ox Bow Press, 1989.

_____. *Metamorphosis of Plants*. Trans. by Anne Marshall and Heinz Grotzke. Wyoming, RI: Biodynamic Literature, 1978.

Gregg, Nicholas. *Primer of Companion Planting*. Wyoming, RI: Biodynamic Literature, 1943.

Grohmann, Gerbert. *The Plant: A Guide to Understanding Its Nature*. 2 vols. Kimberton, PA: Bio-dynamic Farming and Gardening Association, 1989.

Grotzke, Heinz. *Biodynamic Greenhouse Management*. Wyoming, RI: Biodynamic Literature, 1990.

Hagemann, Ernst, Rudolf Steiner, and Harold Jurgen. *World Ether – Elemental Beings – Kingdoms of Nature*. Spring Valley, NY: Mercury Press, 1996.

Hauk, Guenther. *Toward Saving the Honeybee*. London: Rudolf Steiner Press, 2003.

Hauschka, Rudolf. *The Nature of Substance*. London: Rudolf Steiner Press, 1983.

_____. *Nutrition*. London: Rudolf Steiner Press, 1983.

Henderson, Gita, ed. *Biodynamic Perspectives: Farming and Gardening*. Auckland: Bio-Dynamic Farming and Gardening Association in New Zealand, 2001.

Joly, Nicholas. *Wine from Sky to Earth: Growing and Appreciating Biodynamic Wine*. Austin, TX: Acres USA, 1999.

Keyserlingk, Adalbert Graf von. *The Birth of a New Agriculture — Koberwitz 1924*. Forest Row: Temple Lodge Publishing, 1999.

_____. *Developing Biodynamic Agriculture: Reflections on Early Research*. Forest Row, UK: Temple Lodge Publishing, 2000.

Klett, Manfred. *Agriculture as an Art: The Meaning of Man's Work on the Soil*. International Biodynamic Initiative Group Stroud, UK: Biodynamic Agricultural Association, n.d.

_____. *The Biodynamic Compost Preparations and the Biodynamic Preparations as Sense Organs—Talks on Biodynamic Agriculture*. Forest Row, UK: International Biodynamic Initiative Group, 1996.

Klocek, Dennis. *Drawing from the Book of Nature*. Fair Oaks, CA: Rudolf Steiner College Press, 2000.

Koepf, Herbert. *The Biodynamic Farm: Agriculture in the Service of the Earth and Humanity*. Hudson, NY: Anthroposophic Press, 1989.

_____. *Bio-dynamic Sprays*. Kimberton, PA: Bio-dynamic Farming and Gardening Association, 1988.

_____. *What Is Biodynamic Agriculture?* Wyoming, RI: Bio-dynamic Literature, 1976.

Koepf, Herbert & Linda Jolly, eds. *Readings in Goethean Science*. Rhode Island: Biodynamic Farming and Gardening Association, 1978.

Konig, Karl. *Earth and Man*. Wyoming, RI: Biodynamic Literature, 1982.

Kranich, Ernst Michael. *Planetary Influences upon Plants: Cosmological Botany*. Wyoming, RI: Biodynamic Literature, 1986.

Lehrs, Ernst. *Man or Matter: Introduction to a Spiritual Understanding of Nature on the Basis of Goethe's Method of Training and Observation*. London: Rudolf Steiner Press, 1985.

Lovel, Hugh. *A Biodynamic Farm*. Kansas City, MO: Acres U.S.A., 2000.

Pelikan, Wilhelm. *Healing Plants: Insights through Spiritual Science*. Spring Valley, NY: Mercury Press, 1997.

_____. *The Secrets of Metals*. Spring Valley, NY: Anthroposophic Press, 1973.

Pfeiffer, Ehrenfried. *Biodynamic Gardening & Farming*. 3 vols. Spring Valley, NY: Mercury Press, 1983.

_____. *Biodynamics: Three Introductory Articles*. Chester, NY: Bio-Dynamic Farming and Gardening Association, Inc., 1956.

_____. *The Earth's Face and Human Destiny*. Emmaus, PA: Rodale Press, 1947.

_____. *Sensitive Crystallization Processes: A Demonstration of Formative Forces in the Blood*. Spring Valley, NY: Anthroposophic Press, 1975.

_____. *Soil Fertility, Renewal and Preservation: Biodynamic Farming and Gardening*. East Grinstead: Lanthorn Press, 1983.

_____. *Weeds and What They Tell*. Wyoming, RI: Biodynamic Literature, 1981.

Philbrick, John and Helen Philbrick. *Gardening for Health and Nutrition: An Introduction to the Method of Biodynamic Gardening*. Great Barrington: SteinerBooks, 1995.

Podolinsky, Alex. *Active Perception*. Sydney: Gavemer Foundation Publications, 1990.

_____. *Biodynamic Agriculture, Introductory Lectures*. Vols. 1-3. Sydney: Gavemer Foundation Publications, 1985.

Pogacnik, Marco. *Nature Spirits & Elemental Beings*. Findhorn: Findhorn Press, 1997.

_____. *Healing the Heart of the Earth. Restoring the Subtle Levels of Life*. Findhorn: Findhorn Press, 1998.

Poppen, Jeff. *The Barefoot Farmer*. Published by author, 1996.

Pretty, Jules. *Agri-culture: Reconnecting People, Land and Nature*. London: Earthscan Publications, 2002.

Remer, Niclaus. *Laws of Life in Agriculture*. Kimberton, PA: Bio-Dynamic Farming and Gardening Association, 1995.

Sattler, Friedrich and Eckard von Wistinghausen. *Biodynamic Farming Practice*. Cambridge: Cambridge University Press, 1992.

Schilthuis, Willy. *Biodynamic Agriculture*. Hudson, NY: Anthroposophic Press, 1994.

Schwenk, Theodor. *Sensitive Chaos: The Creation of Flowering Forms in Water and Air*. London: Rudolf Steiner Press, 1965.

_____. and Wolfram Schwenk. *Water — The Element of Life*. Hudson, NY: Bell Pond Books, 2003.

Soper, John. *Studying the Agriculture Course*. Stroud: Biodynamic Agricultural Association, n.d.

Steiner, Rudolf. *Agriculture: An Introductory Reader*. Vancouver: Sophia Books, 2004.

_____. *Agriculture—A Course of Eight Lectures*, Trsl. by G. Adams. London: Bio-Dynamic Agricultural Association, 1974.

_____. *Agriculture Course—The Birth of the Biodynamic Method*. Trsl. by G. Adams. Forest Row, UK: Rudolf Steiner Press, 2004.

_____. *(Agriculture) Spiritual Foundations for the Renewal of Agriculture*. Kimberton, PA: Bio-dynamic Farming and Gardening Association, 1993.

_____. *Answers to Questions*. 7 vols. London: Rudolf Steiner Press, 1999-2000.

_____. *Autobiography: Chapters in the Course of My Life, 1861-1907.* Great Barrington, MA: SteinerBooks, 2005.

_____. *Bees.* Hudson, NY: Anthroposophic Press, 1998.

_____. *Calendar of the Soul: 1912-1913.* Great Barrington, MA: SteinerBooks, 2004.

_____. *The Cycle of the Year as Breathing Process of the Earth.* Hudson, NY: Anthroposophic Press, 1984.

_____. *Cosmic Memory — Prehistory of Earth and Man.* Blauvelt, NY: Garber Communications, Inc., 1990.

_____. *Founding a Science of the Spirit.* London: Rudolf Steiner Press, 1999.

_____. *Geographic Medicine—Two Lectures by Rudolf Steiner.* Spring Valley, NY: Mercury Press, 1986.

_____. *Harmony of the Creative Word: The Human Being and the Elemental, Animal, Plant and Mineral Kingdoms.* Forest Row, UK: Rudolf Steiner Press, 2001.

_____. *How to Know Higher Worlds: A Modern Path of Initiation.* Hudson, NY: Anthroposophic Press, 1994.

_____. *Introducing Anthroposophical Medicine.* Hudson, NY: Anthroposophic Press, 1999.

_____. *Intuitive Thinking as a Spiritual Path: A Philosophy of Freedom.* Great Barrington, MA: SteinerBooks, 1995.

_____. *Mystery of the Universe — The Human Being, Image of Creation.* London: Rudolf Steiner Press, 2001.

_____. *Nature's Open Secret.* Great Barrington, MA: SteinerBooks, Anthroposophic Press, 2000.

_____. *Nature Spirits.* London: Rudolf Steiner Press, 1995.

_____. *Nutrition and Stimulants.* Kimberton, PA: Bio-dynamic Farming and Gardening Association, 1991.

_____. *An Outline of Esoteric Science.* Great Barrington, MA: Anthroposophic Press, 1997.

_____. *The Spirit in the Realm of Plants.* Spring Valley, NY: Mercury Press, 1984.

_____. *Spiritual Beings in the Heavenly Bodies and in the Kingdoms of Nature.* Hudson, NY: Anthroposophic Press, 1992.

_____. *Theosophy: An Introduction to the Spiritual Process in Human Life and in the Cosmos.* Great Barrington, MA: Anthroposophic Press, 1994.

_____. *Understanding the Human Being—Selected Writings of Rudolf Steiner.* ed. by Richard Seddon. London: Rudolf Steiner Press, 1993.

_____. *Youth's Search in Nature—A lecture given to the young people of Koberwitz.* Spring Valley, NY: Mercury Press, 1984.

Storl, Wolf Dieter. *Culture and Horticulture: A Philosophy of Gardening.* Wyoming, RI: Biodynamic Literature, 1979.

Suchantke, Andreas. *Eco-Geography: What We See When We Look at Landscapes.* Great Barrington, MA: Lindisfarne, 2001.

Sucher, Willi. *The Living Universe and the New Millennium.* Stourbridge, UK, 1997.

Thun, Maria. *Gardening for Life — The Biodynamic Way.* Glasgow: Hawthorn Press, 2000.

Thun, Maria and Matthias K. Thun. *Biodynamic Sowing and Planting Calendar 2006.* Edinburgh: Floris Books, 2005.

Wachsmuth, Guenther. *The Etheric Formative Forces in Cosmos, Earth and Man.* Vol.1. London: Anthroposophic Publishing Co./ Anthroposophic Press, 1932.

Weirauch, Wolfgang, ed. *Nature Spirits & What They Say. Interviews with Verena Holstein.* Edinburgh: Floris Books, 2004.

Wildfeuer, Sherry, ed. *Stella Natura 2006: Kimberton Hills Agricultural Planting Guide and Calendar.* San Francisco: Bio-dynamic Farming and Gardening Association, 2005.

Wilkes, John. *Flowforms: The Rhythmic Power of Water.* Edinburgh, UK: 2003.

Wistinghausen, et al. *Biodynamic Sprays and Compost Preparations: Directions for Use.* Stroud: Biodynamic Agricultural Association, 2003.

_____. *Biodynamic Sprays and Compost Preparations: Production Methods.* Stroud: Biodynamic Agricultural Association, 2003.

ABOUT RUDOLF STEINER

 Rudolf Steiner (1861–1925) became a respected and well- published scientific, literary, and philosophical scholar, particularly known for his work on Goethe's scientific writings. At the beginning of the twentieth century he began to develop his earlier philosophical principles into an approach to a methodical research of psychological and spiritual phenomena, which he called "anthroposophy," meaning "wisdom of the human being." Steiner wrote over thirty books and gave over 6,000 lectures across Europe. His work has led to innovative and holistic approaches in medicine, philosophy, religion, education (Waldorf schools), special education (the Camphill movement), science, economics, agriculture, architecture, and the arts. Today there are schools, clinics, farms, and other organizations all over the world that are involved in practical work based on his principles. The General Anthroposophical Society, which he founded in 1924, has branches throughout the world.

ABOUT HUGH J. COURTNEY

 After service for several years in the United States Navy, Hugh Courtney obtained a degree as a librarian and spent several more years working in public libraries. In 1975, he discovered the world of biodynamic agriculture and, over a period of years, served an apprenticeship in the making of the biodynamic preparations under the premier preparation maker in the United States, Josephine Porter. Upon her death in 1984, he took over her work in making and distributing the biodynamic preparations, founding the Josephine Porter Institute for Applied Biodynamics, Inc.

* * * * *

The Josephine Porter Institute for Applied Biodynamics was chartered in 1985 as a 501 (c) 3 not for profit corporation for the purpose of research and education on the biodynamic agricultural preparations. It was founded by Hugh Courtney to provide a home for the making of the preparations and to fulfill a lifelong wish expressed by his mentor, Josephine Porter. For the ensuing twenty years, the institute has served the greater biodynamic community in the United States and abroad with quality biodynamic preparations, as well as serving as a primary educational base where those who are interested can gain practical experience in learning to make their own biodynamic preparations.